# EXTREMELY LOUD

# EXTREMELY LOUD

## Sound as a Weapon

JULIETTE VOLCLER

Translated from the French
by Carol Volk

**THE NEW PRESS**

NEW YORK
LONDON

The New Press gratefully acknowledges the Florence Gould Foundation for supporting the publication of this book.

Originally published in France as *Le son comme arme* by Éditions La Découverte, Paris, 2011
Published in the United States by The New Press, New York, 2013
Distributed by Perseus Distribution

All illustrations created by Juliette Volcler.

LIBRARY OF CONGRESS CATALOGING-IN-PUBLICATION DATA

Volcler, Juliette.
[Son comme arme. English]
Extremely loud : sound as a weapon / Juliette Volcler ; translated by Carol Volk.
    pages cm
"Originally published in France as Le son comme arme by Editions La Decouverte, Paris, 2011."
Includes bibliographical references and index.
ISBN 978-1-59558-873-9 (hbk : alk. paper) -- ISBN 978-1-59558-888-3 (e-book) 1. Nonlethal weapons. 2. Noise--Physiological effect--Research. 3. Sound--Physiological effect--Research. 4. Military weapons--Research--History. 5. Psychological torture. 6. Sound--Psychological aspects. I. Volk, Carol, translator. II. Title.
U795.V6513 2013
623.4--dc23
                        2012047132

The New Press publishes books that promote and enrich public discussion and understanding of the issues vital to our democracy and to a more equitable world. These books are made possible by the enthusiasm of our readers; the support of a committed group of donors, large and small; the collaboration of our many partners in the independent media and the not-for-profit sector; booksellers, who often hand-sell New Press books; librarians; and above all by our authors.

www.thenewpress.com

Composition by Bookbright Media
This book was set in Kozuka Gothic Pro and Janson Text

Printed in the United States of America

10  9  8  7  6  5  4  3  2  1

# CONTENTS

Acknowledgments vii

List of Abbreviations ix

Introduction: "We Don't Yet Know What a Sonic
   Body Can Do" 1

1. "Ears Don't Have Lids": Technical Aspects
  of Hearing 7

2. The Death Ray: Infrasounds and Low Frequencies 21

3. "Hit by a Wall of Air": Explosions 43

4. "Totally Cut Off from the Known": Silence
  and Saturation 66

5. "Hell's Bells": Medium-Frequency Sounds 91

6. "No Matter What Your Purpose Is, You Must
  Leave": The Sound of Power 125

Conclusion: "A Passionate Sound Gesture" 151

Notes 155

Index 186

# ACKNOWLEDGMENTS

For their support and advice, a huge thank-you to Adeline Bonnard-Delamar, Alexandra Baraille, Benoît Bories, Claude Ollivier, Élodie Brun-Volcler, Émilien Bernard, François Maliet, Hélène and Jean Volcler, Jean-Baptiste Bernard, Jean-Marc Agrati, Muriel Bernardin, Patrick Watkins, Samantha Lavergnolle, Solenn Moreau, and Yeter Akyaz.

# LIST OF ABBREVIATIONS

| | |
|---|---|
| A3D | Aversive Audible Acoustic Device |
| AFRL | Air Force Research Laboratory (United States) |
| ARDEC | Army Armament Research, Development and Engineering Center (United States) |
| ARL | Army Research Laboratory (United States) |
| ATC | American Technology Corporation (renamed LRAD Corp. in 2010) |
| BSCT | Behavioral science consultation team |
| cd | Candela |
| CIA | Central Intelligence Agency (United States) |
| CNRS | Centre national de la recherche scientifique (National Center for Scientific Research) |
| CSS | Compound Security Systems |
| dB | Decibel |
| SPL | Sound pressure level |
| DARPA | Defense Advanced Research Projects Agency (United States) |
| DMP | Dispositif manuel de protection (manual protection device) |
| DPD | Dynamic pulse detonation |
| FBI | Federal Bureau of Investigation (United States) |
| GSS | General Security Service (Israel) |

| | |
|---|---|
| Hz | Hertz |
| IAF | Israeli Air Force (Israel) |
| IFBG | Improved flash-bang grenade |
| IRA | Irish Republican Army (Ireland) |
| JNLWD | Joint Non-Lethal Weapons Directorate (United States) |
| LMA | Laboratoire de mécanique et d'acoustique du CNRS de Marseille (France) |
| LRAD | Long-range acoustic device |
| NATO | North Atlantic Treaty Organization |
| NDRC | National Defense Research Council |
| NIJ | National Institute of Justice (United States) |
| NIMH | National Institute of Mental Health (United States) |
| NGO | Non-governmental organization |
| ONR | Office of Naval Research (United States) |
| PDT | Pulse Detonation Technology |
| PSS | Plasma sound source |
| RAF | Rote Armee Fraktion (Germany) |
| SADAG | Sequential Arc Discharge Acoustic Generator |
| SARA | Scientific Applications and Research Associates (United States) |
| SERE | Survival, Evasion, Resistance and Escape (United States) |
| SMBI | Stress and Motivated Behavior Institute (United States) |
| TBRL | Target Behavioral Response Laboratory (United States) |
| UN | United Nations |

# INTRODUCTION

# "WE DON'T YET KNOW WHAT
# A SONIC BODY CAN DO"[1]

Considered from a military perspective, the ear is a vulnerable target: you can't close it, you can't choose what it hears, and the sounds that reach it can profoundly alter your psychological or physical state. The second half of the twentieth century saw the development of scientific research on the military and law enforcement uses of sound. The aim was no longer to use sound just to send out an alarm, intimidate the enemy, or rally the troops, but to exploit sound's biological effects, since sound waves—which are nothing more nor less than mechanical vibrations—can harm the ear and the entire body at certain frequencies and certain intensities.

The advantage sound offers to those in power is that it produces the same results as other "non-lethal" weapons, while defusing criticism and confusing the debate. The rapid development of sound weapons or devices, starting in the 2000s, is evidence of a change in law enforcement, in the conception of the public sphere, and in our relationship to those around us. Few popular overviews of this question are currently available, and the snippets of information that circulate often meld fact and fiction. This book offers a critical presentation of the various uses of sound as a weapon from World War II to today.[2] It aims to permit a more general understanding of the stakes of these weapons, and thus to favor the creation of resistance—the flowering of "passionate sound actions," in the words of Escoitar, a Spanish collective that analyzes the acoustic technologies of social control.[3]

## ON THE TIGHTROPE OF SOUND

The use of sound as a weapon is not exactly new. In the 1940s, military loudspeakers appeared on the battlefield as weapons of deception and psychological abuse; Nazi researchers did their utmost to make sound lethal. During the decade that followed, the military and scientists launched sophisticated experiments on sensory manipulation. In the 1960s and 1970s, research was oriented more specifically around the development of infrasonic weapons or deafening grenades. The 1990s placed "non-lethality" at the heart of security doctrine, and interest increased in technologies using medium and high frequencies. In the 2000s, music was used as a means of torture in the "war on terror," while "sound cannons," as the media called them, were deployed at demonstrations, and sound devices proliferated outside commercial establishments or residential properties to distance teenagers or other "undesirables."

Despite all this, there is continual surprise at the existence of acoustic weapons. For more than half a century, articles have been appearing in the military, technical, or even general press presenting these "weapons of the future" as if for the first time. In a world dominated by images, it's as though we need to be reminded constantly of the existence of sound, this background noise, sidekick of the visible. Sound is perpetually forgotten and rediscovered. It's probably also the case that acoustic weapons, with their invisible, untouchable, magical nature, have fueled a literature too passionate to bother separating fact from fiction, too fascinated to distinguish science from the hype promoted by weapons manufacturers or conspiracy theorists. It isn't the weapons in and of themselves that we are constantly rediscovering, but the fact that they actually exist outside of films and novels. Therein lies the task: to try to trace a path between what is unsaid and what is imagined. It's a tightrope walk, a balancing act.

In the 1990s and 2000s, the German physicist Jürgen Altmann devoted several studies to verifying the claims of the weapons industry and the media, and analyzed the real effects of sound on the human body. He notes, for example, that a single weapon is alternatively described as emitting low frequencies and high frequencies, that the effects claimed by another are materially impossible, and that a third, presented as real, never existed. The absence of public information and the dearth of independent experts in this field have provided fertile ground for this convenient confusion; thus, when it comes to sound, one is never entirely safe from illusion, and verification is key. For in this marketplace of fictions, some verifiable outlines have emerged that will profoundly modify our relationship with sound, with space, with power, and with society: a major change, a new arms race, unperturbed by either regulations or public opinion.

Why sound? Why *only* sound? Why draw attention, in reports on torture or inventories of "non-lethal" weapons, only to what concerns audition? Is producing unbearable sounds worse than using lasers that produce intense pain?[4] No. Is the use of music as a means of torture more inhumane than simulated drowning? No. It's not a matter of comparing senses, of taking the side of sound. Nor is it a matter of compiling a catalogue of horrors. Rather, we are attempting to look at history from a different perspective, to lay out the genesis of acoustic repression in order to understand a phenomenon that has been present without our noticing it. Not to complete this history but to redraw it. Sound, a terra incognita to be explored, is a manifestation of the imagination of power.

## WHAT IT'S ABOUT AND WHAT IT'S NOT ABOUT

We spoke of an "overview" of sound as a weapon, but this book is more a subjective montage of words, experiments, and images, a collage of pipe dreams, projects, nightmares, inventions, horrors, failures, testimonies, claims, analyses . . . a long historical cut-up on sound that will reconstruct, in a new but not random fashion, clips of discourse, whether official or industrial, derived from the media or critical analysis.[5] This history will not be exhaustive. To begin with, weapons programs are secret—only those that have been declassified or documented in some manner are accessible. Many of the available sources come from the United States, which dominated the second half of the twentieth century from an industrial standpoint and was particularly active in military research into new weapons, notably those termed "non-lethal." Finally, requests for interviews to obtain additional information from independent organizations, journalists, or researchers were almost all refused or ignored; it may be that sound has not been taken very seriously as a subject of inquiry, or has simply never been envisaged as one.

In reality, the sources are not all that rare, provided you look for them, collect them, and cross-reference them. The discussion will be based on books about sound, weapons, or the human sciences; on some of the numerous patents that the idea of repressive sound has given rise to via many an inventor; on basic scientific, historical, legal, or political articles; on university documents or military academy theses; on reports by non-governmental organizations, governmental organizations, and international institutions; on the military press, never stingy with its inventive metaphors, or on the general press, which is often hungry to relay these metaphors; on the Internet sites of the promoters, manufacturers, and resellers of sound weapons, and on the far fewer sites of those who cover and provide critical information on the issue; on sound or audiovisual works, some of them by individuals who are fascinated by sound as a weapon, others by people who intend to escape this initial hypnotism to propose another relationship to sound and the public sphere.[6]

A note on vocabulary and usage: we prefer the expression "acoustic weapons" or "sound weapons" to "sonic weapons," generally used by video game makers or gamer networks. And unless we are quoting someone who has not made this choice, we use quote marks around the expressions "non-lethal" and "non-lethality": that's the practice followed by British scholar Neil Davison, the author of *"Non-Lethal" Weapons*,[7] a work that synthesizes ten years of study between 1997 and 2007 by the University of Bradford's Non-Lethal Weapons Research Project.[8] As we shall see, "non-lethal" and "less-lethal" are euphemisms at best. Sound weapons do not belong to a well-defined category. Many are the research programs that have tried to use sound to kill. Since the 1990s, acoustic weapons have been classified essentially by their developers among "non-lethal" weapons. But here too there is variation: some authors classify them, along with stun guns or chemicals, as "directed-energy" weapons; others place them in a separate category. Explosives often are not included in any category, even when they are intended to be "non-lethal." If we have chosen to use the phrase "sound as a weapon," it's precisely to avoid these ill-suited categories, and to include in the analysis not only devices but also practices, such as sensory deprivation, sound harassment, or the breaking of the sound barrier.

We will deal with sound here exclusively as a physical phenomenon, applied offensively in a military, police, or other public order context. Therefore we won't discuss sound propaganda,
nor radio or music as a means of political expression. We won't
mention Muzak, except in the discussion of ambient music used
by stores or mass transportation to attract a certain clientele or
create a particular mood; nor will we discuss the most recent
marketing techniques aimed at turning a few notes into an "audio virus" that the consumer can't get out of his or her head.
We will describe only briefly electromagnetic weapons, some of
which use microwaves to produce a sound effect. We will also
leave out the discussion of sound used as a weapon of detection or exploration (sonar, scanners) or authentication (acoustic
signatures). Some of these areas have already been the subject
of detailed publications; each of them deserves a study in itself.

The first chapter aims to be practical, providing the technical keys for understanding the subsequent chapters, which
focus on weapons—may it discourage neither those who are already knowledgeable in acoustics nor those who will be learn-
ing the mechanics of sound for the first time. The book will
then follow the evolution of the range of frequencies, since
each part of the sound spectrum (infrasounds or low frequencies, medium frequencies, high frequencies, blasts, and silence)
corresponds to a specific scientific, military, and commercial
history. The final chapter will offer a more analytical approach,
attempting to understand, to borrow the expression of British
DJ and university professor Steve Goodman in his study on
sonic warfare, "what a sonic body can do"—what it can do in
law enforcement, in dominating the imagination and the urban
experience of our daily lives.[9] In conclusion, we will sketch out
what it can do in the invention of resistance, the emergence of
new imaginary worlds, the new collective spaces to be invented.

# 1

# "EARS DON'T HAVE LIDS": TECHNICAL ASPECTS OF HEARING[1]

Michel Chion, the "listening stroller," notes in the first pages of his treatise on "acoulogie" that "the ear is the only organ that is both external and internal, hence perhaps the particular symbolism we attach to sound, making it a link between various worlds (real and imaginary) and various realms (physical and spiritual)."[2] What follows are some technical elements to help the reader understand what sound is and how it acts on the body. This section is also intended as a glossary, the words in bold being necessary to the comprehension of the chapters that follow.

Sound is a **mechanical vibration** (called acoustic) in a material atmosphere (solid, liquid, or gas) that propagates, like a wave, in time and space[3]: a pulse given in one place is transmitted thanks to the oscillation of molecules or atoms. Thus there is no displacement but rather transmission: the **signal** propagates in this way and is called a **sound wave**. A basic signal is characterized by its frequency, its amplitude, and its speed of propagation in a given medium.

Before expanding on these definitions, we should note that sound signals are perceived by the entire body, but particularly by the ear, the organ responsible for hearing and equilibrium, which absorbs 90 percent of acoustic energy. The auditive apparatus can be divided into three parts: the outer ear, the middle ear, and the inner ear. The **outer ear** is composed of the pinna (also called the auricle) and the ear canal, which protects the internal parts and selects the frequencies that are useful to verbal communication. The **middle ear** includes the eardrum and the chain of three small bones or ossicles: the eardrum receives acoustical vibrations from the air, which cause it to tremble like the skin of a drum, and transmits them to the three bones. While frequencies useful to communication are amplified, basses and strong sounds are filtered. The **internal ear** is composed of the organs of equilibrium (the vestibular apparatus) and of hearing (the cochlea). The spiral-shaped cochlea transmits the vibrations of the ossicles to the brain. It is covered by the basilar membrane, which serves as a base to the organ of Corti, itself composed of nerve receptors equipped with hair cells. When a sound reaches the receptors, these hair cells, distributed throughout the long membrane, begin to move like the keys of a piano, and thus transmit the different frequencies to the brain.[4] The ear functions therefore as a microphone, transforming acoustic energy into electrical energy.

## "HEARING IS TOUCHING FROM A DISTANCE": FREQUENCY[5]

The sound spectrum, which includes all sounds, can be broken into **frequencies** (the number of oscillations of the mechanical vibration in one second), the unit of measurement of which is the hertz (Hz).[6] Rapid vibrations (high frequency) produce a high sound, whereas slower vibrations (low frequency) produce a low sound—the "la" of the tuning fork corresponds, for example, to 440 Hz.[7] One characteristic that interests the military (in particular) is that the human ear hears only a part of the sound spectrum. Generally we don't hear sounds below

20 Hz (**infrasounds**) or above 20,000 Hz (**ultrasounds**)—the sound exists, but we don't hear it, or not much.[8] Between the two is what's called the **audible range**: low frequencies (deep sounds) are between 20 and 200 Hz, medium frequencies (medium sounds) fall between 200 Hz and 2,000 Hz, and high frequencies (high-pitched sounds) between 2,000 and 20,000 Hz. Elephants can hear sounds starting at 0.1 Hz; dogs can hear up to 35,000 Hz, and bats up to 100,000 Hz.

When a sound is made up solely of its **fundamental frequency**, its base frequency, we speak of **pure tone**. Such sound, which is usually of electronic origin (for example, the frequency of an alarm), is aggressive to the ear.[9] It is represented in the form of a sinusoidal waveform (the same movement repeating periodically; see figure 1).

In music, as in nature, we find mainly **complex tones**, which is to say composed not only of their fundamental frequency but also of **harmonic frequencies**, which are multiples of the fundamental frequency. For example, if the "la" base frequency is 440 Hz, it will have harmonic frequencies at 880 HZ (440 × 2), 1,320 HZ (440 × 3), 1,760 Hz (440 × 4), and so on. The fundamental frequency is the lowest and the one with the greatest breadth, so we hear it with the greatest intensity. The **timbre** of an instrument is determined by the combination of the fundamental frequency and the harmonics that it emits, the relative intensity of which varies, as does its evolution in time: that is why the "la" of a piano sounds different from the same "la" played on the violin.[10] Finally, we speak of octaves to designate the interval between two base frequencies, one of which is the double of the other (for example, the "la" at 880 Hz and the " la" at 1,760 Hz). A complex tone can be represented in the form of a period wave that is non-sinusoidal (see figure 2).

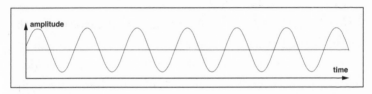

**Fig. 1** Sinusoidal wave ("la" generated by a computer: pure sound at 440 Hz)

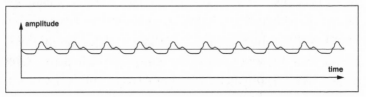

**Fig. 2** Non-sinusoidal periodic wave (repetition of a syllable)

A complex sound can also be non-sinusoidal and non-periodic, if it doesn't repeat the same vibration regularly (see figure 3).

**Noise** is an aleatory or irregular wave (see figure 4). In acoustics, therefore, noise is not an unpleasant or unaesthetic sound, but a continuous signal in which no particular frequency can be distinguished. Over the course of the twentieth century, contemporary music, notably bruitism, questioned, reinterpreted, and deconstructed these definitions in their relationship to the concept of aesthetics. What we call **white noise** is a noise containing all the frequencies at the same intensity (as when a television is not properly adjusted).

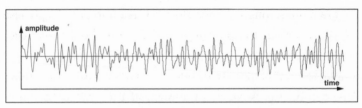

**Fig. 3** Non-periodic, non-sinusoidal wave (sound background in a bar)

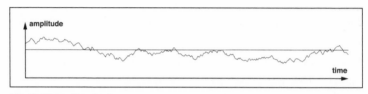

**Fig. 4** Noise (hissing)

The capacity to hear low and high frequencies varies from person to person, according to age and health. The ear is far more sensitive to frequencies between 1,000 Hz and 4,000 Hz, which are useful for oral communication, than to low frequencies. Another notable characteristic of our bodies is that it is not only the ear that reacts to sound: in proximity to a loudspeaker or a **subwoofer**,[11] which sends low notes at a high volume, our intestines vibrate. We call this the **extra-auditive effects** of sound. Our body perceives a part of the infrasounds and the ultrasounds that are inaudible to the ear. A deaf person may dance perfectly well, therefore, since he or she may feel the vibrations of the sounds in his or her body.

A final clarification concerning frequency: every physical body (object or organism) has its own particular frequency, or **resonance frequency**, which causes maximal vibration in the object (harmonic frequencies also cause vibration, but less so). Sound touches every body around it and is transformed into mechanical energy, into vibration—like the surface of water when you throw a stone. When the frequency of sound coincides with the resonance frequency of an object, the object vibrates more strongly. When Bianca Castafiore, the "Milanese Nightingale" in the Tintin series of comic books, manages to break glass with her voice, it is because the frequency she emits is the same as the resonance frequency of glass: under the combined effects of the frequency (the note), the intensity (or volume), and the length (how long the singer holds the note), glass vibrates with increasing force to the point of breaking. The same holds true for a bridge, which can be destroyed by a weak gust of wind if the wind's resonance frequency corresponds to that of the bridge.

Frequencies inherent to the human body vary according to the organ:

> For vertical vibratory excitation[12] of a standing or sitting human body, below 2 Hz the body moves as a whole. Above, amplification by resonances occurs with frequencies depending on body parts, individuals, and posture. A main resonance is at about 5 Hz where the greatest discomfort is caused; sometimes the head moves strongest at about 4 Hz. The voice may warble at 10 to 20 Hz, and eye resonances within the head may be responsible for blurred vision between 15 and 60 Hz. In-phase movement of all organs in the abdominal cavity with consequent variation of the lung volume and chest wall is responsible for the resonance at 4–6 Hz.[13]

But we often get too excited about the vibrational potential of infrasounds, forgetting one important thing: that low-frequency acoustic waves, which are not unidirectional, apply the same pressure to the whole body. The result is that "air pressure variations impinging on the human body produce some vibration, but due to the large impedance mismatch nearly all energy is reflected."[14] Therefore "intense levels of low-frequency noise would be necessary to achieve the same level of discomfort resulting from low-frequency vibration applied to the body via mechanical contact."[15] The most sensitive parts of the body at these frequencies are those that contain large volumes of air: the ears and the lungs. The stronger vibration of the thoracic cavity comes into play between 40 Hz and 60 Hz.[16] This particular sensitivity of the human body to certain frequencies is exploited in a positive manner by certain medical techniques or in some spiritual practices (mantras). It is also taken into account in the manufacture of cars, as infrasonic frequencies emitted by a vehicle potentially cause varying levels of discomfort in the driver or among the passengers (notably nausea and fatigue).

## "THE SOUND OBJECT IS NOT A STATE OF MIND": AMPLITUDE[17]

The Bianca Castafiore character plays not only on sound frequencies but also on **volume**. Since the technical texts that we cite refer regularly to **sound pressure levels**, we should point out, without getting into details, that the intensity (measured in watts per square meter), pressure (measured in pascals), and amplitude (measured in decibels) are different manners of evaluating the same phenomenon, the **sound level**. We will refer here to the decibel SPL (sound pressure level), abbreviated as dB.[18] Zero dB corresponds to the minimum the human ear can hear: it's the threshold of audibility, not absolute silence. Ants emit stridulations, spiders cry, and beetles make noise as they move about inside tree trunks, as evidenced by the recording and amplification[19] work of audio-naturalists Boris Jollivet and David Dunn,[20] but because these acoustic expressions are situated below the threshold of audibility, we don't hear them. Whispering reaches around 20 dB, a washing machine 50 dB, a normal conversation 60 dB, a busy road 80 dB, and a plane taking off 140 dB.

The human ear, thanks to the contraction of muscles of the middle ear, is capable of protecting itself from sounds that are too strong, but it needs silence to recuperate; the stronger the intensity of the exposure, the more extensive the silence needed for recuperation to occur. After a loud explosion, the ear experiences a temporary displacement of its threshold, or a **temporary threshold shift** (TTS). Recuperation can take a few minutes or a few hours, but in cases of harsh contrast, of prolonged or repeated exposure, the trauma causes a permanent shift of the threshold, or a lasting loss of hearing.[21] We should note, finally, that so-called **perception deafness** is due to nerve lesions in the inner ear (destruction of hair cells, notably), while **transmission deafness** derives from lesions of the bone or obstructions in the external or middle ear.

A very precise study by Jürgen Altmann, "Acoustic Weapons, a Prospective Assessment: Sources, Propagation and Effects of Strong Sound," analyzes the organic effects of strong sound as a function of the frequency, the amplitude, the distance, and the length of exposure.[22] We'll come back to his study in a more detailed manner in reviewing the different uses of sound as a weapon, but provide here the scale of amplitude he establishes by way of example: the ear can be subject to damage, without our even necessarily being aware of it, starting at 85 dB, the **discomfort threshold** is reached at around 120 dB, and the **pain threshold** is around 140 dB.[23] Two things should be specified, however: first, this scale varies greatly according to the type of sound and the person; second, we will use here the numbers provided by Altmann, but the pain threshold is usually considered to be 120 dB.

The difference between the physicist's higher threshold and the commonly accepted value lies no doubt in the use made of these numbers.[24] Altmann performs scientific work aimed at verifying and debunking the numerous claims, many of them fanciful, on the effects of sound, while governmental health authorities, associations that fight noise, and other actors devise regulations aimed to protect hearing or promote acoustic moderation. Exposure to noise in a professional setting is the subject of a European directive promulgated in 2003, then transposed into national laws.[25] We will work here with Altmann's numbers not to raise the threshold, but for three reasons: first, because he furnishes one of the very rare independent studies on the organic effects of sound and on acoustic weapons, constructing his synthesis through a critical analysis of numerous sources; second, with an eye to coherence, given that we will base our arguments regularly on his evaluations; finally, because his rigor sheds light on the claims made by certain promoters of acoustic weapons, as well as on the denial of certain effects made by others.

Between 140 dB and 170 dB, various effects may appear that are generally temporary, among them respiratory problems, chest pressure, excess salivation, blurred vision (for low frequencies), nausea, a feeling of heat, tingling (for high frequencies), dizziness, **tinnitus**,[26] a loss of hearing, headaches, fatigue, and an accelerated heart rate. Above 160 dB, eardrums rupture. A shock wave (in other words, an explosion) above 200 dB can tear the lungs, and one above 210 dB can cause fatal internal hemorrhaging. That said, Altmann rules out certain effects claimed by the manufacturers of acoustic weapons and their users (armies, police), such as vomiting, bowel spasms, or uncontrolled defecation.[27]

At the same intensity, certain frequencies are perceived by the human ear as stronger than others. For example, at 40 dB, a frequency of 1,500 Hz (to take that of the long-range acoustic device or LRAD, a powerful loudspeaker that we

will discuss later) will be perceived as stronger than a frequency of 15,000 Hz. Another notion of importance is that
decibels follow a logarithmic scale: when the power of a sound
is doubled, it translates into an augmentation of 3 dB (and
not a doubling of decibels), while multiplying the power by
one hundred translates into an increase of 20 dB. If the operator of the LRAD turns it up from 120 dB to 123 dB, he is
doubling the level of the sound. Similarly with two different
sound sources: if you hear a scooter arriving (at 80 dB) on a
busy street (where the sound level is at 80 dB), the combined
level will be 83 dB (and not 80 dB or 160 dB). Finally, strong
sounds mask weak sounds, and low frequencies tend to cover
higher frequencies: contrary to what happens in the visual
field, in which objects cohabit, sounds overlap and mix with
one another, which is referred to as the **masking effect**.

## AN ANCIENT POND,
## THE FROG LEAPS:
## THE SILVER PLOP AND GURGLE OF WATER: PROPAGATION[28]

The speed of sound's propagation depends on the medium
through which it passes: the denser the medium, the closer
the atoms or molecules, and the faster the propagation. In
the air, sound propagates at 340 meters per second (m/s), or
1,224 kilometers per hour, with variations according to temperature or altitude. Its speed is thus far slower than that of
light, which can reach 300,000 kilometers per second. That
is why we see a flash of lightning before hearing the thunder, even though they're produced simultaneously. A plane
"breaks the sound barrier" when it moves more quickly than
the sound it produces, thereby "breaking" its own acoustic
wave and provoking a **shock wave**. Sudden changes in acous-

tic pressure produce a **bang** (explosion) and a **blast** (a wind effect). The crack of a whip is also due to the breaking of the sound barrier by its loop. In water, sound travels at 1,500 m/s, and through steel at around 5,900 m/s—hence the recurring scenes in Westerns of cowboys and Indians putting an ear to the rails to detect a distant train. In the soft tissue of the body the speed of propagation is equivalent to that of the aquatic environment, while in the bones it is about 4,000 m/s.

The medium also has an effect on the length and nature of the propagation of the wave. In front of an obstacle, sound is diffracted—it travels around it—but the **diffraction** varies according to the amplitude of the wave, its frequency, and the size of the obstacle: if the obstacle is large, a low frequency can circumvent it, while a high frequency can only get around a small obstacle. Why doesn't a sound wave propagate indefinitely? Because it is absorbed by matter, which diminishes its amplitude, whether gradually or abruptly. The insulating materials in an **anechoic chamber** not only block all external sounds but also absorb internal sounds: no sooner is a sound produced than it is extinguished. In an **echo chamber**, on the contrary, the absorption is delayed because the sound signal is reflected off every surface: therefore you hear the initial signal (a clap of hands in the middle of the room, for example) and its reverberations bouncing off the four walls, the ceiling, and the floor. This, in simple terms, is the phenomenon of the **echo**.

The echo is produced in a natural and continuous manner, but if it is weak, the human ear considers the initial signal and its reverberations as a single sound. If the difference between the initial signal and its reverberations becomes stronger, the echo is perceived consciously. In a concert hall, various absorption techniques are used on the walls and ceilings in order to limit reverberations. In ancient amphitheaters and in medieval churches, "sounding vases" placed in the structure played a double role: that of absorption (attenuating the reverberations of sound) and that of amplification of the voices of the actors or singers (the frequency of the resonance of the vases, around 200 Hz, was perfectly adapted to male voices, which at the time had a monopoly on the stage).[29] On the contrary, **sonar** uses the reverberating properties of the sound wave: it sends a signal and, based on eventual echos, can determine if there is an obstacle (which reflects the signal) and how far away it is. This so-called echolocation technique, or biosonar, is also used by bats, which emit ultrasounds in order to orient themselves in space.

Three specific phenomena of propagation can occur and are exploited by the makers of acoustic weapons, and more generally by sound industries: in-phase, out-of-phase, and stationary waves. We speak of waves that are **in phase** when waves of the same frequency propagate at the same time, attaining their maximums and their minimums simultaneously: the amplitudes add together to produce a stronger sound (see figures 5 and 6).

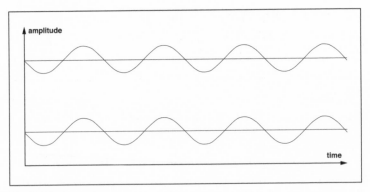

**Fig. 5** Waves in phase

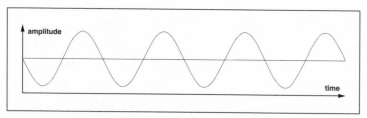

**Fig. 6** Wave resulting from the superposition of waves in phase

On the other hand, when waves are **out of phase**, the first reaches its maximum at the moment when the second attains its minimum: the amplitudes cancel each other out, producing an inaudible or weak sound (see figures 7 and 8). In anti-noise headsets, a small integrated microphone captures exterior noise, while loudspeakers send it into phased opposition to cancel it.

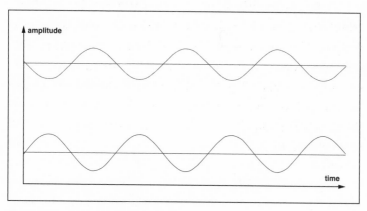

**Fig. 7** Waves in phased opposition

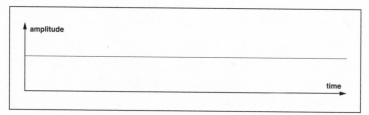

**Fig. 8** Wave resulting from phased opposition

A **stationary wave** is created when a sinusoidal wave is propagated in a closed environment. It goes forward and back on the same axis in the medium, one direction sometimes in synch with the other, and sometimes in phased opposition. The intensity of the sound varies according to the site.

Finally, waves propagate differently according to their frequencies; we speak of the directionality of sound. Waves spread in a spherical manner around their source, but higher frequencies propagate in a more directional fashion and are more rapidly absorbed or reflected by their medium. Low frequencies propagate in every direction and travel far, as they are less impeded by obstacles: infrasounds produced by volcanos can therefore travel several kilometers, their intensity attenuating gradually. And if we can better hear the bass notes of the neighbor's stereo system than its high notes, it's because the walls, floors, and ceiling reflect or absorb high-pitched sounds, while the low vibrations propagate (passing from aerial transmission to transmission through solids) or diffract. Loudspeakers, like microphones, also have a directionality: some are omnidirectional, others favor sound heading toward or coming from a particular direction. The ear, an organ of warning, perceives where sound comes from: a sound wave coming from our left first reaches the left ear, then, with a slight delay and already attenuated, the right ear. The auditory system reconstructs it as a single and same sound, but the minute difference allows us to detect the source.[30]

The musicologist Joachim Ernst-Berendt concludes: "Unconsciously we always react to noise like Stone Age beings. At that time a loud sound almost always signified danger. . . . [The word *alarm*] comes from the Italian . . . *all'arme*, a call to arms. When we hear noise, we are constantly—but unconsciously—'called to arms.'"[31]

# 2

# THE DEATH RAY:
# INFRASOUNDS AND LOW FREQUENCIES

Even though research and development was later oriented mainly toward medium- and high-frequency weapons, it was the domain of infrasounds and low frequencies that first interested researchers and the military. It is also the area that generated the greatest flights of fancy. Understandably so: infrasonic frequencies are able to enter into resonance with the frequencies of the human body. Infrasounds are omnipresent in nature, at an intensity that is not dangerous for us: the waves of the ocean, waterfalls, volcanos, and earthquakes emit infrasonic frequencies, which animal ears perceive better than we do, since the human ear, designed to prioritize the frequencies useful to communication, filters out low frequencies. The reputation of the Mediterranean mistral, the "wind that makes you crazy," derives perhaps from, in addition to its unpredictability, the fact that it emits a high amplitude of frequencies between 5 Hz and 20 Hz.[1] According to Galen, a second-century Greek doctor, Roman courts "deliver more lenient judgments during the season of sirocco, for it is known as a time when even normal people behave in strange ways."[2] Industrialization, with its parade of machines and motors, multiplied the number of infrasounds present in daily life—along with the nuisances linked to what we have called "noise pollution." For example, virulent campaigns against wind power emerged as these turbines sprouted, due in particular to the low frequencies they emit.

At the level of 80 dB to 100 dB, low frequencies disturb sleep, and in an industrial environment disturbances are noted at amplitudes below 60 dB.[3] In 2003, the British Department for Environment, Food and Rural Affairs (DEFRA) produced a report on low frequencies that emphasized the disparate conclusions of studies evaluating the impact of these frequencies on the human body: some claim an absence of significant effects, while others found evidence of dizziness, imbalance, extreme disturbance, disorientation, apathy, nausea, intestinal spasms, and abdominal or heart irregularities. It mentions as well the fact that workers exposed for fifteen minutes to industrial infrasounds of 5 Hz to 10 Hz at a level of 100 dB to 135 dB report fatigue, apathy, depression, ear pressure, loss of concentration, confusion, and vibration of internal organs.[4]

Independent observers are more limited in their analysis of the effects of low frequencies on the body. Altmann makes no mention of either intestinal spasms, disorientation, apathy, confusion, or depression and notes that certain effects such as dizziness and nausea are wrongly attributed to infrasounds, whereas they are mainly brought on by low or medium frequencies.[5] Even the threshold of pain is attained at higher levels with infrasounds than with higher frequencies—for a frequency of 2 Hz, it is considered to be 162 dB (and no longer 140 dB), for example.[6] Before Altmann, British physicist Nick Broner had already noted that "a review of the effects of low frequency noise indicates that the effects are similar to those of higher frequency noise, that noise in the 20–100 Hz range is much more significant than infrasound at similar sound pressure levels and that the possible danger due to infrasound has been much overrated."[7] The problem with studies such as DEFRA's lies in the fact that, except under experimental conditions, a person rarely experiences an environment that includes *only* infrasounds (or *only* ultrasounds): it is difficult, therefore, to distinguish between the effects due to audible frequencies and those brought about by inaudible frequencies.

The influence of low frequencies on the organism has been studied most notably by NASA in the 1960s, in order to evaluate the negative impact on Apollo astronauts.[8] One NASA researcher, George C. Mohr, found that, based on experiments with volunteers, sounds below 100 Hz become disturbing starting at 150 dB:

> At such levels these effects include moderate chest wall vibration, a sensation of gagging, a blurring of vision and amplitude of modulation of speech. While such exposure was within the voluntary range of the subjects taking part in the experiments, they were not considered pleasant and post-exposure fatigue was present for 24 hours after repeated testing. There were no reports, however, of permanent injury in this work. The voluntary tolerance limit seems to be around 150–155 dB, with headaches being reported at 50 Hz, pain on swallowing and giddiness occurring both at 66 Hz and 73 Hz. At 100 Hz mild nausea, giddiness and flushing were produced.[9]

Loss of hearing isn't mentioned. And in any case, the legend of the "brown note" (or "brown noise"), alleged to induce an involuntary bowel movement through infrasonic magic, is not confirmed anywhere, so demonstrators can head out with no fear on this score. One final note before tackling the history of infrasonic weapons: it is necessary to distinguish between the effects noted after a period of long or repeated exposure to low frequencies—in a professional environment, for example—and the immediate effects a weapon would need to produce based on these same frequencies.

## "ITS PERFORMANCE WAS VERY NEARLY LETHAL": THE GAVREAU MYTH[10]

The harmful potential of infrasounds was discovered by chance by a French acoustician, Vladimir Gavreau, who ran the Laboratoire de mécanique et d'acoustique (LMA) of the CNRS (Centre national de la recherche scientifique) in Marseille. The date of his discovery varies according to the source: many mention 1957,[11] but Pierre Liénard, a former member of the board of the LMA, believes it was 1967 and supports this, in his *Petite histoire de l'acoustique*, with an interview with Gavreau that appeared in *L'Aurore* that year.[12] As for Gavreau himself, in an article published in January 1968 by *Science Journal*, he states that the experiment took place "four years ago."[13] This uncertain chronology is only part of what makes Gavreau a learned mix of reality and fiction. In his study on "sonic warfare," Steve Goodman calls him "a key hyperstitional figure"[14]—"hyperstition" being a new concept to designate fictional entities becoming real, or rather, in the case of Gavreau, a reality transformed by fiction. If the concept as such brings nothing new and perhaps is overly complacent, Goodman points out rightly the permanent interconnection between science and the imagination that characterizes infrasounds and acoustic weapons in general. It is not the least of their appeal, as we shall later see.

But let's get back to the very real facts that gave rise to the myth. Gavreau had observed in himself and his colleagues physiological reactions that were as severe as they were inexplicable:

Everything started a few months ago: I was working in my office with an engineer, Mr. Albert Calaora, when we suddenly fell seriously ill. We had the impression that our heads were going to burst, and it soon became unbearable. Intrigued, we wanted to get to the bottom of it, and soon we detected the presence of very low-level, thus inaudible, infrasounds. Shortly thereafter, we found where it was coming from: a giant ventilation system that had been installed that very day in a neighboring building.[15]

The defective machine was emitting a frequency of about 7 Hz, which, amplified by the air vents where it had been installed, had an unpleasant effect on the body:

Our eardrums were periodically compressed—a process that was both exceedingly painful and potentially dangerous. The intensity of the infrasound was so strong that everything vibrated: tables, glassware on the tables, liquids in various vessels; also, curious patterns appeared on the surface of liquids. Even the needle of an ordinary barometer oscillated. But all our microphones indicated "a perfect silence."[16]

Gavreau then decided to concentrate his research on infrasounds, their effects on the human body, and infrasonic weapons that might be derived from them: "It is based on this phenomenon and this incident that the idea struck us to produce infrasounds to annoy our friends."[17] Specifically, Gavreau sought to perfect a machine that would not just generate infrasounds, but *direct* them, countering the physical properties of low frequencies, which are diffused in a spherical fashion around their source. Gavreau developed several prototypes. He reproduced the frequency of 7 Hz via an organ pipe 24 meters long embedded in concrete.[18] He also created what he called an "acoustic laser," made up of seventy-four pipes of the same length, activated in synchronicity.[19] Gavreau stated that the effects of the frequency of 7 Hz "were certainly unpleasant, producing a characteristic throbbing in the head and making the most simple intellectual task impossible," and he wondered if "this was due to the particular frequency involved? Seven hertz is the medium frequency of the 'alpha waves' of the brain, which corresponds to an absence of any kind of intellectual work."[20]

He then built an instrument that he called an "acoustic gun,"[21] but which in no way resembled a gun; it consisted of a contraption somewhere between an organ pipe and a Levavasseur whistle,[22] all of it embedded at a right angle in a 200-kilogram block of cement. The result, according to Gavreau and his colleague Henri Saul, who were equipped with hearing protection, was a frequency of 196 Hz emitted at 160 dB, which produced "a painful 'resonance' in our bodies—everything inside us seemed to vibrate when we spoke or moved." It was "very nearly lethal," according to Gavreau. "Presumably, if the test had lasted longer than five minutes internal hemorrhage would have occurred. The feeling of internal resonance disappeared after about three hours."[23]

Finally Gavreau and his team developed a gigantic whistle, 1.5 meters in diameter, which, paired with organ pipes, emitted a frequency of 37 Hz: "The walls of our laboratory began to vibrate and little crevices appeared!"[24] Liénard concludes in his *Petite histoire de l'acoustique*: "Enclosed in a room, with no one inside, the device at maximum strength shook the houses in the neighborhood, caused the dogs to howl, and provoked such a commotion that the people threatened to attack this *nut house*, which then lowered the intensity of the sound. No other infrasound experiments were ever conducted at the CNRS in Marseille."[25] Thus concluded the history of what might have been an easy-to-handle directive infrasonic weapon.

Legend has it that the scientists got very sick and took several days to recover: this was the start of a series of sensationalist articles on the "lethal death rays," which were neither rays nor lethal. Gavreau's account should be taken with a grain of salt, if we are to believe Altmann: "Note that today scientists at the same institute have some doubts about the conclusions drawn by Gavreau on the effects of infrasounds, because his experiments and observations have not been replicated and confirmed under accurate experimental conditions."[26] Journalists were not alone in propagating and amplifying Gavreau's myth: one of the most zealous promoters was the writer William Burroughs. During an interview with guitarist Jimmy Page published in the music magazine *Crawdaddy* in 1973, Burroughs gushed with enthusiasm: "[Gavreau] had an infrasound installation that he could turn on and kill everything within five miles. It can also knock down walls and break windows. But it kills by setting up vibrations within the body. Well, what I was wondering was, whether rhythmical music at sort of the borderline of infrasound could be used to produce rhythms in the audience."[27] The industrial-music band Throbbing Gristle would fuel the legend in the 1970s by using infrasounds (and some ultrasounds) in their concerts—this time, the miraculous frequencies were credited with producing orgasms as well as involuntary defecation in the audience.

Altmann is more circumspect, notes no effect on intellectual processes, and sees possible danger of death only in the case of a high-amplitude shock wave, and not at low frequencies.[28] On the other hand, he notes severe respiratory problems starting at 150 dB at frequencies between 50 Hz and 100 Hz, as well as scars on the tympanic membrane of submariners who had long exposure to intense infrasounds.[29] Previously researchers Stanley Harris and Daniel Johnson had exposed men to frequencies of 7 Hz at 142 dB for fifteen minutes without observing "lowered performance," dizziness, or disorientation.[30] This is enough to invite skepticism with regard to patents claiming "subliminal acoustic manipulation of nervous systems" based on infrasonic frequencies.[31] Broner, on the other hand, considers that an infrasonic weapon could eventually become fatal from 174 dB on—but specifies:

> Thus, the sound power required even for an ideal system at 250 meters is on the order of a thousand times that generated by a Saturn V rocket on lift-off. A further problem with such a device would be the large source size required. Bryan and Tempest have calculated that the source would have to be 1,100 m in diameter to have good directional properties for the radiation.

He concludes: "It seems, therefore, that military use of infrasound is not practical, and that the report which suggested that 'the Trompette Marseillaise with which Gavreau thought he could re-enact the feat of Joshua before the walls of Jericho is only a beginning' was wishful thinking."[32]

## "REDUCE THEM TO QUIVERING DIARRHOEIC MESSES": MILITARY DREAMS[33]

Gavreau nonetheless sparked several decades of scientific and military research on the organic effects of infrasounds, which were used as tools to detect enemy artillery starting with World War I.[34] Today this function of probing and localization still remains their primary military usage. As for their impact on human beings, the objective of the first research led by NASA in the 1960s was to evaluate and limit their potentially noxious effects. The second half of the twentieth century saw several programs on infrasounds, primarily carried out by the United States: this time their aim was to develop infrasounds as weapons, lethal or "non-lethal." With rare exceptions, researchers ended by concluding that this offensive usage was impossible or inefficient. The history of these efforts provides the general framework for the development of acoustic weapons.

### From the 1960s to the 1980s: Efforts and Fantasies

Reports mention research on the use of low frequencies as weapons for Great Britain and Japan during World War II, with no further details provided.[35] Then in 1969, in *Riot Control*, a book about the repression of uprisings, U.S. colonel Rex Applegate briefly mentions tests on "supersonic and subsonic" weapons and specifies that they were put aside due to their cost, the difficulty of handling them, and their lack of precision.[36]

In 1973, an article in *New Scientist* reports on the use by the British army of the "squawk box" to disperse riots in Northern Ireland:

> The squawk box emits two marginally different frequencies, almost out of the audible range, through separate speakers. These combine in the ear to produce two other frequencies—one the sum of the two component frequencies and the other the difference between them. Thus two speakers emitting, say 16,000 Hz and 16,002 Hz produce

a high frequency in the ear of 32,002 Hz and a low beat of only 2 Hz.[37]

Its usage was envisaged in combination with another prototype, the photic driver, a stroboscope that aimed to use intermittent flashes of light to produce epileptic seizures, nausea, speech difficulties, or fainting, and which was developed by London-based Allen International.[38] Despite all these details, the existence of the squawk box appears doubtful: either it never got beyond the experimental phase, or there might have been confusion with another weapon. In *"Non-Lethal" Weapons*, Neil Davison points out that "the MOD [the British Ministry of Defense] denied the existence of the weapon and instead acknowledged that they had acquired the HPS-1 audible sound system."[39] Developed for the U.S. Army, this powerful loudspeaker was used during the Vietnam War, as we shall later see. Altmann doesn't rule out the possibility of producing an infrasound thanks to the combination of two very high frequencies.[40] However, like Broner, he suggests that an extremely strong, and in his view unrealistic, sonar intensity would be required for such an infrasound to induce auditive pain and dizziness.[41]

In 1978, Hungary was worried about the development of "infrasonic weapons of mass destruction" and presented a study to the United Nations that argued for interdicting the production and development of such weapons, which it claimed could have "psychotropic effects," causing "throat spasms" or "injuring human internal organs."[42] If these warnings were not based on any precise facts, they were nonetheless symptomatic of the fear and fascination engendered by infrasounds at a time when states envisaged the invention of a lethal usage.

The creation of an "infrasonic bullet" by Russia was mentioned in the press in the mid-1990s—Janet Morris, a military consultant for the United States, returning from a Moscow-based testing center for non-lethal devices, declared, "We saw a ten hertz acoustic generator which would be used to deliver a pulse about the size of a baseball that could knock you down or more . . . kicked up to lethal level. . . . Range of the hand-held acoustic generator is hundreds of yards."[43] Altmann doubts the accuracy of the account and the reality of this weapon, given that the wavelength corresponding to a frequency of 10 Hz would be quite a bit larger (34 meters) than a baseball, and that the reach of this "bullet" doesn't seem credible; he suggests the possibility of confusion with other technologies, such as the vortex or plasma, which we will later discuss.[44] The Soviet Union and then Russia did conduct research into the biological effects of infrasounds in the 1980s and 1990s, but no offensive application is documented.[45]

In 1997, a Chinese journal of military medicine noted the existence of "an infrasound weapon [that] had already been developed and tested and that the device was adjustable, to cause controllable amounts of disorientation, nausea, vomiting, and incontinence."[46] Not only was no complementary information given, but the claims strongly resemble those refuted by Altmann. William Arkin, a military consultant specializing in humanitarian issues, indicated that same year, "No acoustic weapons are currently known to exist, and even the truth behind rumored Soviet-Russian experiments and breakthroughs is doubted."[47]

### The U.S. Research Program: 1990s–2000

In the early 1990s, the United States made a concerted effort to launch research into a "low-collateral-damage munitions program," led by the Army Armament Research, Development and Engineering Center (ARDEC), based at Picatinny Arsenal in New Jersey. Twenty government laboratories and military command centers, as well as a half dozen private companies, were brought into the program, the principal industrial contractor being the company Scientific Applications and Research Associates (SARA).[48]

The United States sought to add four types of acoustic weapons to its arsenal: a short-range weapon, a rifle type of weapon or an "acoustic water cannon" for peacekeeping missions or hostage rescue; a longer-range weapon that could be mounted on a helicopter or vehicle for missions to maintain order, invasions, or zone protection; an "air-dropped acoustic mine," to neutralize soldiers or workers in military or industrial zones; and fixed "acoustic fences" to protect military or nuclear zones as well as borders.[49] Optimism was the order of the day: "the beams," its proponents said, "would be capable of producing waves of energy that rain in like kicks to the stomach." High-intensity sound would attack soldiers, "liquify their bowels and reduce them to quivering diarrhoeic messes."[50] Beginning in 1991, SARA was given various contracts by several governmental agencies: ARDEC, the Defense Advanced Research Projects Agency (DARPA), and the Department of Energy.

SARA worked on the conception of several weapons, first and foremost ultrasound, which was judged the most promising. We therefore find in the catalogue of its projects the infrasonic pulser, a generator of infrasonic pulses emitting frequencies between 1 Hz and 17 Hz; a potentially lethal low-frequency "acoustic bullet," emitted by a parabolic generator between 1 and 2 meters in diameter; combustion or compressed-air sirens, emitting infrasounds or audible frequencies; a "non-lethal" weapon for helicopters, sending a sound signal from more than 2 kilometers away between 100 Hz and 10 kHz; an "acoustic beam" of frequencies between 20 Hz and 340 Hz with an amplitude of 120 dB to 130 dB to prevent access to arsenals of weapons of mass destruction; the pulsed periodic stimuli generator, which produced harmful sound pulses starting at 110 dB; and a "sonic firehose," or, less lyrically, a directed loudspeaker.[51]

While some of these weapons, such as the infrasonic pulser, the siren, and the directed loudspeaker, reached the prototype stage, independent researchers remained highly skeptical about the claims made in SARA's press statements, which spread "some of the myths described [in the popular media]": "the ability to interfere with the nervous system, causing disorientation, or inducing a passive state within the targeted subject," lethargy, vomiting, intestinal pain, trauma, hemorrhages, and spasms.[52] This skepticism certainly seems to have been well advised, given the limited posterity of these inventions, which today have all disappeared from SARA's website. But in the meantime the company did develop the concept of the "high power acoustic beam weapon," which we find in other forms in medium and high frequencies.[53]

In 1996, the United States created the Joint Non-Lethal Weapons Directorate (JNLWD) in order to coordinate research and development on "non-lethality," a new strategic priority. Acoustic weapons were the focus of special attention, and a specific program was launched, the Non-Lethal Acoustic Weapons Program, directed by the army in conjunction with the Air Force Research Laboratory (AFRL).[54] The goal was to develop "acoustic generators that could be used to demonstrate extra-aural effects in people."[55] It was in this context that the Army Research Laboratory (ARL) built the Sequential Arc Discharge Acoustic Generator (SADAG), producing sound pulses, by means of high-voltage electrical shocks, of up to 20 Hz in frequency and 165 dB in amplitude.[56] But the tests performed by AFRL were not convincing. In a report issued in October 2000, the laboratory frankly concluded that "none of the four devices tested would have obvious utility as a non-lethal weapon."[57] In the testing lab was the SADAG, a combustion-driven acoustic siren, and a compressed-air-driven siren (the last two constructed by SARA), as well as the Gayl blaster (named for its inventor, Franz Gayl, then a major in the Marine Corps).[58] While the SADAG has some effect on the behavior of hogs, it did not affect that of monkeys but did render them half deaf,[59] which was far from the desired outcome for a "non-lethal" weapon capable of inducing "relatively reversible effects."[60]

DARPA and the National Institute of Justice (NIJ) at that time financed AFRL to undertake research into infrasounds and low frequencies and to "demonstrat[e] the utility of ultra-low frequency sound as an incapacitation technology suitable for hostage rescue scenarios."[61] AFRL built the Infrasound Test Device and the Mobile Acoustic System, a mobile infrasonic generator equipped with a 17-meter loudspeaker: "The only potentially useful effect was an apparent panic response shown by rhesus monkeys . . . at greater than 160 dB . . . but the required intensity of infrasound would probably be hard to sustain in the 'battlespace' at any distance."[62] AFRL later placed pigs, then monkeys, in a reverberating chamber equipped with subwoofers and a fixed wall producing stationary infrasound waves at 10 Hz, 12 Hz, 15 Hz, and 20 Hz. The monkeys and the pigs showed "minimal impact," even at 145 dB. In the final report, the experimenters concluded again that the effects of infrasounds and audible sounds on the organism are "grossly overstated" and that "it seems unlikely that high-intensity acoustic energy in the infrasonic or low-frequency range will provide a device suitable to be used to facilitate hostage rescue."[63] And, as in the report on SADAG, they "would not suggest the need for further extensive experimentation in this area of research."[64]

At the end of the 1990s, a now-defunct company, Synetics Corporation, received government authorization to develop an infrasonic ray that could be used as "a non-lethal crowd control [and] non-lethal self defense units for police and personal use."[65] In 2001, its Internet site mentioned development for the Picatinny Arsenal of a device very close to the phantasmagoric squawk box: the Difference Acoustic Wave Generation System, "a high power acoustic generation system that would produce two narrow ultrasonic acoustic beams . . . This effectively allows infrasonic or low frequency acoustic energy to be precisely directed to a small target area."[66] But the prototype seems to have sunk with the company. Beginning in September 1999, the program of research on acoustic weapons was officially halted, in view of its limited results: "Specifically, the program needed to show a prototype device that could produce a reliable, repeatable bio-effect with sufficiently high infrasound amplitude at a minimum specified range. In August 1999, the Acoustics Program Manager communicated that the effort could not meet the transition criteria."[67]

In 1996, the British Ministry of Defense reached similar conclusions in a feasibility study affirming the impossibility of producing operational acoustic weapons.[68] Wyle Laboratories had tested low frequencies on volunteers, in order to evaluate the possibility of preventing access to a bunker thanks to high-intensity sounds. An electromechanical valve controlled the expulsion of compressed air into a room, thus producing various frequencies; the most effective were those between 63 Hz and 100 Hz. Their effect nonetheless did not convince the report authors, who concluded that infrasonic frequencies would probably be more useful but that the intensity required would be difficult to attain.[69]

Research on infrasounds nonetheless continued, mainly at the initiative of two laboratories founded by ARDEC: the Target Behavioral Response Laboratory (TBRL), which studied the possibility of variable-effect weapons (from "non-lethality" to lethality) based on acoustic technologies or directed energy, and the Stress and Motivated Behavior Institute (SMBI), located at New Jersey Medical School and specializing in neurobehavioral research.[70] ARDEC, which calls itself the army's "Center for Lethality," financed SMBI beginning in 2003 to test the effects of infrasonic frequencies on rats.[71] The results, reported on SMBI's website, were not spectacular: at best, the rats demonstrated slower reaction times or lost their balance more quickly on an exercise wheel, and aside from that, the infrasounds "do not cause noticeable effects."[72] But this time the pursuit of experimental research was recommended.[73]

A few applications also seem to have potential in the aquatic medium: what is true in the air is not necessarily true in the water, where the body does not react the same and sound propagates differently. Contrary to aerial acoustic waves, which mainly reverberate off the human body, acoustic waves in an aquatic medium pass essentially directly from the water to the body.[74] This particularity makes possible the development of "non-lethal" underwater weapons, which aim to keep swimmers and divers at a distance. As with low frequencies in the air, research rests on analyses initially conducted with non-offensive goals in mind: that of protecting fetuses, immersed in amniotic fluid, or divers exposed to sonar, underwater explosions, or other aquatic tools.[75]

In a 2002 publication on the Non-Lethal Swimmer Neutralization Study, conducted by the Applied Research Laboratories at the University of Texas at Austin, two prototypes were noted in the realm of low frequencies: the Harbor Subsurface Protection System by Lockheed Martin and an "airgun-based sound source" by Weidlinger Associates. The first was supposed to be able to project a frequency of 32 Hz at an amplitude of 200 dB for 6 meters, producing "intolerable discomfort, including visual distortions, mask and sinus vibrations and thorax vibrations." The latter could "send out a 4–10 Hz high intensity sound," and claims were made that "low frequency resonance of the visceral organs in the thorax and abdomen will be excited, producing discomfort." The manufacturers nonetheless provided no references or evidence.[76] The study did not consider that the claims should be taken at face value, but nonetheless concluded that "the most promising long-term solution for a non-lethal diver threat response is the development of a low frequency sound source [particularly within the range of 20-100 Hz] designed specifically to produce signals likely to cause discomfort in human divers."[77]

Are low or very low frequencies therefore definitively unsuitable for all police or military use? As we have seen, the history of the research is more than just anecdotal—it shows the difficulty to this day of constructing weapons, whether lethal or "non-lethal," that are portable, easy to handle, use a reasonable amount of energy, and are capable of targeting a precise object while sparing the weapon's operator.[78] We shall see that explosive devices, which emit generally low frequencies while playing on other properties, have had greater success. But in the meantime, the main infrasonic device turns out to be the "infrasound fish fence," currently used to discourage fish from coming too close to industrial turbines during their migration.[79] As for humans: in more spectacular than offensive fashion, some U.S. police have equipped their vehicles with Rumblers, sirens that use low and medium frequencies (from 182 to 400 Hz, according to the website of the Federal Signal Corporation),[80] causing cars and people nearby to vibrate and thus improving the persuasive capacity of law enforcement.[81]

## "THE GHOST IN THE MACHINE": MYSTERY INSTRUMENT[82]

If military applications based on low frequencies seem extremely limited, certain possibilities—also very limited—exist in terms of entertainment. Burroughs took some liberties with Gavreau's experiment but saw correctly the possible exploitation of low frequencies by industrial culture—and also, no doubt less consciously, by religion.

An experiment similar to Gavreau's, although less extreme, was reported by British researchers at the University of Coventry, Vic Tandy and Tony R. Lawrence, in an article entitled "The Ghost in the Machine." In 1998, Tandy was working in a laboratory that had the reputation for being "haunted." Although skeptical, the researcher observed several strange phenomena, including a general feeling of depression that would strike people (including himself) in a specific room, along with shivering, hair standing up on the neck, and the impression of seeing gray shapes moving stealthily in the room: "It would not be unreasonable to suggest I was terrified."[83] The ghost in reality was yet again a ventilation system that emitted an infrasound. In this case it was a stationary wave of 19 Hz, which, sufficiently amplified by the casing that surrounded it, caused eyeballs to vibrate (hence the gray apparitions and other vision troubles) and induced respiratory problems as well as a vague sense of oppression. Tandy repeated the experiment in a fourteenth-century cave in which "ghosts" had been seen: after recording and analyzing low frequencies, he again discovered a peak of 19 Hz. This time, the amplification occurred thanks to a long corridor leading to the cave, while the source remains unknown.[84]

This infrasonic "anxiety" was analyzed by the anthropologist Donald Tuzin, who published his observations in a 1984 article entitled "Miraculous Voices." He looked at the use of sound in the ritual practices of the Ilahita Arapesh in New Guinea and mentions two instruments in particular, the bullroarer[85] and a long amplifying pipe, a hollow bamboo stalk 4 meters long and 7 centimeters in diameter, one end of which is inserted in a drum. These "secret" instruments are central to worship in the Tambaran cult and produce sounds so "weirdly disturbing" that they are attributed to ritual spirits: the bullroarer is the "voice of Lefin," a spirit who spits drums and gongs, and the pipes are "the voice of Nggwal," the greatest of Tambaran spirits. Tuzin himself was profoundly impressed by these ritual sounds: "Amplifying pipes and bullroarers are alike in their ability to generate an experience so unusual as to validate belief in supernatural reality."[86]

Subsequently analyzing the reasons for this effect, he offered the following hypothesis:

> The bizarre effects infrasonic waves have on our perceptual apparatus are traceable to . . . their potential for producing high-intensity sound pressure in the absence of perceived sound. . . . Under normal circumstances a

loud sound gets our attention. This occurs because the reticular activating system, from its location in the brainstem, monitors all ascending sensory pathways and, in the event of extraordinary sensory input, projects to various parts of the brain (including the neocortex) impulses that bring the organism into a state of alertness. In the case of infrasonic-wave exposure, the reticular activating system would presumably function in the same manner—with the result, however, that the interpretive centers in the neocortex, for want of additional or specific sensory clues, would lack an object on which to direct their attention and in terms of which to coordinate a cognitive or sensorimotor response. The consequence might well be the disruption of sensorimotor functions, accompanied by a more or less pronounced feeling of anxiety, dread, and disquieting vulnerability.[87]

Tuzin suggests that the "quiet before the storm," the impression of strangeness in the instants preceding an earthquake, for example, may be due to the same causes: the inability of the brain to interpret acoustic vibrations (infrasounds produced by an earthquake propagate several kilometers) that reach a person only through audition.[88] The bullroarer and the amplifying pipes have the capacity to "trigger those neural mechanisms underlying the so-called religious experience."[89]

In 2003, Sarah Angliss, a British composer and engineer, launched an "infrasonic" project: inspired by Tandy and Lawrence, she brought together a team of acousticians, psychologists, musicians, and artists to lead an experiment on the emotional impact of infrasounds.[90] To do this, an infrasonic cannon was built using an extra-long-stroke subwoofer in a sewer pipe that was around 6 meters long, emitting at a frequency of 17 Hz. Two contemporary music concerts were then organized in the Purcell Room in London; during some of these sessions, the infrasound generator was activated at a weak level from 6 dB to 8 dB, masked by the music. A questionnaire was distributed to the 750 audience members, in order to measure their "emotional reactions." According to the team's results, the presence of infrasounds increased "strange feelings" by 22 percent: goose bumps, rapid heartbeat, sadness, excitement, and other spiritual sensations. Church organs, the longest pipes of which can produce very low if not infrasonic frequencies, may thus play a role in stimulating the exaltation of worshippers.

The film industry has also exploited infrasonic frequencies or low frequencies. In 1974, Universal Pictures released the film *Earthquake*, which for the first time used the Sensurround process, developed by the makers of Cerwin-Vega speakers:

> Cerwin-Vega acoustical engineers began work on the first Sensurround feature, *Earthquake*, by analyzing tape recordings of the 1971 Sylmar quake to determine the frequency distribution of the infrasound. The Sensurround system was able to propagate bass fundamentals down to 16 Hz at 100 decibels. The theory was that this would be structurally safe since buildings resonate at 6 Hz. Even so, during the first test at the Grauman's Chinese Theater in Hollywood a net had to be installed to catch the ornate ceiling moldings which began dropping to the floor.[91]

In 2002, the director Gaspar Noé added an infrabass of 27 Hz to the first half hour of the soundtrack of *Irréversible*, in order to heighten viewers' sense of fear: "You can have a physical reaction to the movie, too. . . . So for the first half of the movie you feel weird, you could show just a cat drinking milk, and it'd be scary, and you wouldn't know why but it's because of this infra-wave beneath."[92]

After more than a half century of speculation and many military and industrial research programs, the real influence of existing infrasonic machines on the body seems indeed to be limited to some quivering and chills. This is not to minimize the impact of undesired low frequencies in a personal or professional environment, nor to deny the potentially harmful effects of infrasound at very strong intensities, but simply to take account of what exists in the domain that interests us, sound as a weapon. It is difficult to get around the fact that given the current state of knowledge, low frequencies essentially serve to sell subwoofers and science fiction films—with no regrets that the horrors promised never materialized. The "infrasonic boom" seems to have done just fine without them. An associate of Sarah Angliss borrowed one of her infrasound generators and put it on display at an art installation with a label warning visitors about its "powers." Visitors complained of strange effects, and one of them even asked that the generator be turned off. But it had never been turned on.[93]

# 3

# "HIT BY A WALL OF AIR": EXPLOSIONS[1]

Explosives, detonation engines, and other air cannons are not generally classified among acoustic weapons. One reason is that the frequencies emitted by these weapons are hard to identify: they are more noise than sound. Sometimes, even, it is best to speak of wind, since at very strong amplitudes, with no clear transition, we go from sound waves to variations in atmospheric pressure, which are no longer, strictly speaking, acoustic. The composer Curtis Roads calls them "perisonic intensities": at the periphery of sound, and perilous.[2] Moreover, whether incidentally or intentionally, these weapons have "combined effects," for the sound doesn't act alone but is accompanied by a shock wave, a flash, or a projection of fragments or chemical substances.

Based on their position at the frontiers of acoustics, these devices shed a unique light on the use of sound as a weapon. Most of the energy they produce is situated in the low or very low frequencies, and they emit at a sound level that is far from harmless.[3] It is true that this is not particular to these devices, as all explosions share these characteristics: firecrackers reach an amplitude of between 130 dB and 190 dB, and military artillery between 160 dB and 190 dB.[4] That said, it is not a matter of evaluating the sound impact of firearms or of establishing a history of explosives as a whole, but rather of concentrating on a few particular weapons for which the acoustic power is a consciously calculated effect and not a simple by-product—those that seek to retain only the sound element of an explosion, or only the wind effect of a sound. The development of these weapons cannot in fact be separated from research on the lethal or "non-lethal" effects of sound: they emerge from it or they benefit from it, to the point that their history and that of low frequencies are intertwined, with the distinction that the makers and users of these weapons, precisely because the weapons are not strictly acoustic, cross boundaries respected in sound weapons. Being less defined, these weapons are also potentially more dangerous.

At a weak intensity, sound pulses have the same impact on the ear as continuous sound. But at high intensity, the brusque change of pressure connected to the shock wave affects the body more violently than a continuous sound after only a few exposures, or even just one. A temporary loss of hearing can result after an exposure of 2 milliseconds at 140 dB. At 185 dB, a three-second exposure is enough to cause a tearing of the eardrum, a permanent loss of hearing at certain frequencies, serious tinnitus, and hemorrhages of the trachea and lungs, the latter being particularly sensitive to shock waves. The vestibular system is also vulnerable to these pulses: problems of equilibrium and dizziness have been noted, for example, among soldiers rendered deaf by detonations. A few milliseconds of exposure at 200 dB brings a fissure of the lungs, and at 210 dB internal hemorrhaging can cause death by suffocation or obstruction of the blood vessels.[5] Various studies have observed that a strong and sudden sound induces a muscular reflex, which can cause a simple drop in performance or, in certain cases, lead to cardiac arrest. When it is produced underwater, such a noise can engender panic reactions in divers, with fatal consequences at times.[6]

## "LIKE HAVING A BUCKET OF ICE WATER THROWN INTO YOUR CHEST": VORTEX AND PLASMA[7]

The first documented occurrence of the development of a weapon using abrupt changes in pressure goes back to the *Wunderwaffen*, literally "wonder weapons"—the name given by the Ministry of Propaganda to the experimental weapons program of the Third Reich. Shortly before the Nazi surrender, American major general Leslie E. Simon was sent with others to Germany to lead an inquiry into the program. He would publish his observations in 1947 under the title *German Research in World War II*.[8]

He mentions in particular that under the authority of Albert Speer, then minister of armaments, a research center situated near Lofer, Austria, worked "to duplicate in miniature the effects of tornados" thanks to a vortex cannon.[9] The vortex is a (natural or artificial) phenomenon that takes the form of a whirlwind in which the moving particles (air, water) wrap in a spiral around a zone of low pressure.[10] To produce a vortex in a controlled manner, Dr. Zippermeyer, who was responsible for this research, used a mortar set in the ground, which launched a projectile filled with carbon powder and a weak explosive charge. Once in the air, according to the scientist, the powder explodes and a vortex is created if the projectile is moving at a speed of at least several hundred meters per second. The idea was to be able to "remove the wings" of planes, which would be unable to sustain the resulting pressure differential. Simon indicates that "he achieved a considerable vortex effect," but he doesn't mention any use of the cannon other than an experimental one.[11] Another prototype was developed by a company in Stuttgart: a "wind gun" aiming to shoot a "plug of air" at an airplane to destroy it.[12] A model of this cannon projecting air by means of a mix of oxygen and hydrogen was found at the test center in Hillersleben. The Germans working on-site announced that the device could "break one-inch boards at a range of 200 meters but it produced no appreciable effect on aircraft at normal ranges." It failed to ensure the anti-aircraft defense of a bridge on the Elbe.[13]

Vortex weapons were the subject of major research in the twentieth century, not only in Germany but also in the Soviet Union and in the United States.[14] It is not just the kinetic energy of the vortex, its power of impact, that interests researchers and military folks, but also the fact that its centrifugal force allows it to transport other particles. No weapon seems to have moved beyond the prototype stage. During World War II, the American inventor Thomas Shelton worked on resolving the problem of the unpredictability of combat gases, which a strong breeze can send back toward those who launch them. He developed a device that propels a vortex of noxious gas, which can thereby transport the poison over long distances. The prototype "sent a 45 cm smoke ring a distance of 50 meters with an 'eerie howling sound.'"[15] It would never be used.

In the early 1970s, the United States showed particular interest in developing "vortex rings and wind-generation machines" for "crowd and mob control," but no known result emerged.[16] In 1996, Dr. Andrew Wortman of a company called Istar proposed the development of a "vortex ring generator" with the same goal, but the army did not pursue the research "because it required fielding an entirely new system, and the trend in the Army was to reduce weight and logistic costs."[17] In 1997, ARL and ARDEC began to recycle and proposed adding a "kit" to the MK19-3 grenade launcher, which would provide "a means of quickly converting the Navy MK19-3 automatic 40-mm grenade launcher between lethal and non-lethal modes of operation" and allow it to shoot not only grenades but also gas vortices transporting chemical products. In the end, it was concluded "the kit enables the weapon to apply flash, concussion, vortex ring impacts, marker dyes, and malodorous pulses onto a target at frequencies approaching the resonance of human body parts," but "gaps in technology . . . inhibit fielding."[18] In terms of developing "non-lethal" weapons, that same year the JNLWD launched a dedicated program, the Vortex Ring Gun Program, still with ARL. It was an ambitious project: "The Vortex Ring Gun (VRG) program will design, build, and successfully demonstrate the capability to produce combustion-driven, ring vortices that will deter and disorient hostile individuals or crowds." Once again, a combination of the vortex with other effects was envisaged:

> Applications could include an ability to mark an individual or object with a fluorescing dye at a distance; delivery of an incapacitating agent at a distance;[19] delivery of aerosol at a distance (a chemical to corrode, lock, or otherwise disable an automobile); or temporarily introducing a smoke screen or obscuring agent.[20]

But the research was not "satisfying": a stop was put to the program in 1998 due to the "unpredictable vortices and limits on effective range."[21]

Nonetheless, the enthusiasm didn't abate. Five years later, British researchers Neil Davison and Nick Lewer reported:

An acoustic technology receiving considerable R&D attention is the vortex generator. . . . At the 2nd European Symposium on Non-Lethal Weapons in 2003 several groups presented on this topic. These included papers by The Defence Science and Technology Laboratory (DSTL) of the U.K. Ministry of Defence on "Initial Simulations of a Single Shot Vortex Gun," Bauman Moscow State Technical University reported research on "Application of Vortex Technologies for Crowd Control," and the Fraunhofer Institute of Chemical Technology (ICT) presented a paper entitled "Impulse Transport by Propagating Vortex Rings—Simulation and Experiment."[22]

In 2004, Canada showed a certain interest in vortex weapons in a report on "non-lethal" weapons.[23] And in 2006, even though the research had been interrupted officially eight years prior in the United States, SARA's website still boasted the merits of its vortex weapon: "A supersonic vortex of air hits its target at about half the speed of sound with enough force to knock them off balance. The vortex feels like having a bucket of ice water thrown into your chest." Despite its capabilities, the weapon does not seem to have been used and has since disappeared from the company's website.

## "I FELT LIKE I WAS IN THE MIDDLE OF A BOMB": EXPLOSIONS AND SHOCK WAVES[24]

In the context of the Third Reich's program of "wonder weapons," the center in Lofer, Austria, was also working to use audible sound as a weapon—but it was the pressure of the sound wave that interested the Germans more than frequency. Dr. Richard Wallauscheck created parabolic reflectors, the last and "most successful" of which was 3.2 meters in diameter. A short tube connected to the back of an antenna served as a "combustion chamber or sound generator." It produced explosions by means of a mix of methane and oxygen. This first shock wave then reflected back from the extremity of the tube, producing a second explosion in the chamber. The device emitted detonations at a frequency of "800 to 1500 impulses per second" and at an amplitude of 134 dB at 60 meters.

The Germans, and Simon after them, credited the weapon with all sorts of capabilities: although "no physiological experiments had been conducted . . . it was estimated that at such a pressure it would take from 30 to 40 seconds to kill a man. At greater ranges, perhaps up to 300 meters, the effect, although not lethal, would be very painful and would probably disable a man for an appreciable length of time. . . . The general opinion was that the military value of such a device was limited, to say the least, owing chiefly to the lack of range."[25] The weapon was nonetheless passed down to posterity thanks to the Belgian writer and illustrator Hergé, who created a faithful reproduction in Tintin's *The Calculus Affair* (*L'Affaire Tournesol*) based on a photo published by Major General Leslie Simon in *Secret Weapons of the Third Reich*. Hergé took greater liberties from a technical point of view, having Professor Tournesol invent a weapon emitting ultrasounds, which logically would have implied a markedly different configuration of the device.

### Research in the United States

In the 1950s, the Central Intelligence Agency (CIA) launched Project MKUltra, a vast program aiming to study "behavior modification," to which we will return at length in our discussion of research on sensory deprivation. The highly controversial program was the subject of a Senate hearing in 1977. The report of this hearing mentions notably that Subproject 54, conducted in 1955, aimed "to cause brain concussions and amnesia by using weapons or sound waves to strike individuals without giving and without leaving any clear physical marks."[26] A declassified document attached to the report mentions the installation of an explosives zone in order to test two procedures. With regard to the first, it notes that a "single blast pressure wave propagated in the air must have considerable intensity in order to produce brain concussion." Concerning the second, "concentration of the sound-field at some remote point could be effected with acoustical lenses and reflectors." All this would occur with great discretion: "The blast duration would be on the order of a tenth of a second. Masking of a noise of this duration should not be too difficult."[27] These lofty objectives seem to have remained at the experimental stage: during the testimony of Admiral Stansfield Turner, who denied implementation of the project, Senator Richard Schweiker stated that according to his information, the Office of Naval Research (ONR) struggled for at least a year to create "the perfect concussion."[28]

In the late 1980s, the U.S. Army invented the "Voice of God," classified it as top secret, and then put it away in the closet. That, at least, is the legend, as told by science journalist Justin Mullins:[29] "At an undisclosed military research facility hidden in the New Mexico desert . . . researchers working with high-power laser weapons discovered that they could create a glowing ball of fire in the sky by crossing the beams of two powerful infrared lasers." The plasma obtained was so intense that "even the smallest shock waves sound like firecrackers." The researchers then "created a stream of shock waves that merged together to form a continuous loud hiss

or, depending on the frequency of the pulsations, a crackle."
Next the team discovered that "by modulating the frequency
and intensity of the hissing sound, they could create a voice-
like effect." The military envisioned its use as a psychological
weapon during the second Gulf War (1990–91), but "for some
reason these plans were never realized." End of story. Mullins went off in search of confirmation from various scientists
and returned empty-handed: "After many telephone calls and
several interviews, the consensus seemed to be that the Voice
of God was possible in the lab but very difficult to produce in
the sky."

At the close of the next decade, the United States became
interested more prosaically in "crowd control" and "area denial" by means of "devices that generate acoustic energy by
repetitive combustion or detonation."[30] As with SADAG, developed by ARL, the tests were not convincing. In the early
2000s, the Applied Research Laboratories at the University of
Texas experimented with comparable weapons, but this time
for underwater use: the "spark gap sound sources" or "plasma
sound source" (PSS). The principle was the following:

> A charge is stored in a large, high voltage capacitor bank,
> and when the PSS is fired, all the stored energy is released
> in an arc across electrodes in the water. The underwa-
> ter spark discharge creates a high-pressure plasma/vapor
> bubble in water. The expansion and collapse of this bubble
> generates an acoustic signature similar to the signatures
> generated by air guns, 354 underwater explosions, and
> combustible sources.[31]

Essentially used to sound the marine depths, in the manner of sonar, the PSS was considered by the military to be

"an attractive candidate for swimmer deterrence because it can be pulsed randomly or repetitively, which would allow it to be used for infrasound or startle response bioeffects."[32] The PSS offered every advantage: not only "the majority of its sound energy occurs in the 20–200 Hz regions, where data indicate that the most likely lung and vestibular effects will occur," but "in addition to producing sound, it also produces a bright flash." And with that "the PSS produces no shrapnel or projectiles, is electronically activated, [and] has an adjustable, calibrated power level."[33] A prototype, called the Underwater Deterrent Security System, was built in 1992 by GT Devices to protect ports, but "neither the Defense Nuclear Agency nor the Navy chose to fund advanced development of the GT Devices system."[34] This didn't discourage researchers, who confided anecdotally that certain divers found the PSS "very unpleasant," while others "refuse to be in the water, even on the surface, when the PSS is operating." In short, the researchers recommended "that the PSS be evaluated, through animal and eventual human testing, as a swimmer deterrent device."[35]

In 1998, a company called Primex Physica International worked on a prototype of an acoustic blaster, which combined four sources of pulsed detonations to generate acoustic pressure of 165 dB at 15 meters, for area denial or crowd control.[36] No further mention is made of the weapon. In 2007, the U.S. Army announced that the company Stellar Photonics had developed dynamic pulse detonation (DPD) technology, allowing for the creation of explosions in the air. Journalist David Hambling, who specialized in the weapons industry, indicated that "a short but intense laser pulse creates a ball of plasma, and a second laser pulse generates a supersonic shockwave within the plasma to generate a bright flash and a loud bang"; the weapon had a range of 100 meters.[37] The combining of DPD with a directional loudspeaker such as the LRAD, which we shall describe later on, was envisaged in the Plasma Acoustic Shield System—the whole providing an effect "much like that of an Mk 141 stun grenade," according to the developer/builder, "but with highly agile speed-of-light delivery and exceptional accuracy and adjustability."[38] The objective was to "stun and disorient" an enemy. But the weapon, tested at the Picatinny Arsenal, lacked power: "it would take several minutes to burn through a piece of paper using the laser."[39] In 2009, despairing of its lethal possibilities, the army passed the project to JNLWD, which specialized in the development of "non-lethal" weapons, and which refused to further fund the project. The conclusion: "Unless the device can be made cheap, effective and reliable, it will be discarded as a gimmick."[40]

### Cannons and Shock Waves in Israel

There exists nonetheless another device, less futuristic and in real live use, midway between the vortex launcher (which it resembles visually) and the acoustic detonator (which is close in principle): the detonation cannon, which generates a shrapnel-free explosion by means of gas detonators. Like a certain number of "non-lethal" technologies, the weapon was first developed to control animals:[41] the "scare cannon" and other "automatic scarecrows" are used in many countries to keep away birds and small rodents.[42] They create propane explosions that generate one or several detonations below 120 dB. Among the producers of these, notably, is the Israeli manufacturer PDT Agro,[43] which in the late 2000s had the idea of transforming its Thunder Generator for birds into a shock wave cannon for humans.[44] Another company, Armytec, which is engaged in military research and development, was tasked by the Israeli minister of defense to commercialize military and paramilitary versions.[45] Credit for the invention goes to an Israeli researcher of Russian origin, Igor Fridman, president of PDT Agro.

An article in the U.S. military press dated January 2010 specifies:

> Using a patented process involving Pulse Detonation Technology (PDT), the system feeds the gas-air mixture into one or more so-called impulse chambers or cannon barrels, where the burning fuel detonates and intensifies in force as it travels through the chamber, exiting in a rapid-fire succession of high-velocity shock bursts. . . . According to company data, the system generates 60 to 100 bursts per minute, each traveling at about 2,000 meters per second and lasting up to 300 milliseconds. . . . The resulting shocks create a double deterrent to rioters and potential intruders, developers here say, by the extreme air pressure and sonic boom effect generated once the mixture propagates and expands through the air.[46]

A cannon with joints, moreover, allows for shooting at a 90-degree angle, which can thus circumvent a wall or any other obstacle, according to Shlomo Tabak of Armytec.[47]

The available information is limited to that furnished by the manufacturer or the army and relayed by the press; it should therefore be taken with a grain of salt. The reach of the weapon is understood to be 50 meters, and according to Fridman, "anyone within 30 to 50 meters from the cannon will feel like he's standing in front of a firing squad." At fewer than 10 meters, the weapon can cause irreversible wounds or be lethal.[48] The idea of killing with a shock wave is not impossible, as we saw earlier, as long as the wave attains an amplitude at least equal to 210 dB—a rather improbable level that no weapon of this type has claimed.[49] The intensity of the shock wave cannon is not mentioned by the manufacturer, which simply states that it is "suitable for operation within hostile and friendly populations"—a very democratic weapon.[50]

The Israeli army itself has devised another sort of large shock wave cannon: by flying at low altitude, military combat jets of the Israeli Air Force (IAF) break the sound barrier. Over four days in September 2005, shortly after the "disengagement" of the Israeli army from the Gaza Strip, twenty-nine supersonic booms were logged, mainly at night or at dawn.[51] Journalist Gideon Levy recalls the history of this practice, terming it "collective punishment" and "mass indiscriminate intimidation":

> In 1969, two Phantoms were sent to sow fear in the skies of Cairo. A year later, Phantoms from the Patishim ("Hammers") squadron did this in the skies of Damascus. This is how a bully demonstrates his strength. Over the years, we also used this method in the skies of Lebanon. But our enemies have never known the type of wholesale booms like those of recent weeks in Gaza.[52]

Thousands of windows were broken, and cracks appeared in some walls. The Palestinians "liken the sound to an earthquake or huge bomb. They describe the effect as being hit by a wall of air that is painful to the ears, sometimes causing nosebleeds and 'leaving you shaking inside.'"[53] A Gaza businessman said: "Sometimes you hear the rockets the Israelis fire but this was different. I felt like I was in the middle of a bomb. When I ran out the door I thought I might find the rest of the street was gone."[54] The inhabitants of a nearby kibbutz confirmed that "the blasts occur every 20 minutes to an hour, and it sounds as though a bomb is falling right into the house."[55]

In November 2005, two human rights organizations, Physicians for Human Rights Israel and the Gaza Community Mental Health Program, petitioned the Israeli Supreme Court, arguing that the procedure was contrary to international law and dangerous to human health. The UN Relief and Works Agency for Palestine Refugees noted that "a majority of the patients seen at its clinics as a result of the sonic booms were under 16 and suffering from symptoms such as anxiety attacks, bedwetting, muscle spasms, temporary loss of hearing and breathing difficulties."[56] Eyad El Sarraj, the director of the Gaza Community Mental Health Program, declared: "When it happens night after night you become exhausted. You get a heightened sense of alert, waiting continuously for it to happen. People suffer hypertension, fatigue, sleeplessness."[57] The Palestinian minister of health reported that the booms brought on miscarriages and heart problems, and the UN, citing panic attacks in children, demanded that they stop.[58]

The attorney representing the state noted during arguments before the Supreme Court that the booms aimed to "disrupt terror activities, engender fear among terrorists planning to attempt to fire rockets, deceive, create disinformation and a sense of threat and confusion among terrorists concerning the extent of Israel Defense Forces operations— their nature and specific locations."[59] As for the Israeli government, it argues that "sound bombs are preferable to real ones."[60] One year later, the Supreme Court still had not ruled on the humanitarian complaints.[61] In a report dated January 2008, the UN mentions that the practice continued in the occupied territories.[62] A few months later, the UN Special Rapporteur on Palestinian human rights, Richard Falk, noted, "There is widespread deafness among the people of Gaza that is blamed on the frequent sonic booms."[63] In December of the same year, the IAF broke the sound barrier on several occasions in the sky over Lebanon.[64] In January 2009, the World Bank reported that a water purification station in Beit Lahiya, in the north of the Gaza Strip, was weakened: "The integrity of the lake structure is endangered by the potential impact of nearby explosions and sonic booms and possible heavy rain." It warned that "failure of the lake structure would put about 10,000 residents of the surrounding area in danger of drowning and spark a wider environmental and public health disaster," and recommended "a wide no-fire zone should be secured around the lake."[65]

## "OVERWHELM FOUR OF THE SENSES": INCAPACITATING GRENADES[66]

In the late 1960s, U.S. colonel Rex Applegate suggested in his book on riot control that police use the military's practice grenades, since these weapons, which are blanks, produce a bright light and a strong sound: "The item is small and easy to carry, and would be very useful in crowd and mob dispersion, especially at night."[67] Other "non-lethal" grenades already existed at the time, without being referred to as such: these "offensive grenades" explode with little shrapnel and produce a deafening sound. They were used not only by the army but also for law enforcement, such as in France in May 1968, or during a demonstration by fishermen in Rennes on

February 4, 1994. Less dangerous than "defensive grenades," which are specifically made to kill, they can nonetheless severely mutilate and even be deadly in certain cases. The idea that developed at the end of the 1960s and during the 1970s was therefore to come up with a new type of grenade, distinct from offensive and defensive models.

The French company Alsetex, a supplier to French law enforcement, submitted a patent in 1970 for an "explosive grenade with no shrapnel." It explained that "in certain circumstances, such as peacekeeping missions, one may be called upon to use explosive charges, the psychological effects of a detonation being very effective. Obviously, however, exploitation of these psychological effects must not be accompanied by the danger of shrapnel wounds."[68] It mentioned that "the use of a small explosive charge, while effective psychologically due to its detonation, is also highly effective for spraying and dispersing an additional product. For example, one can thereby obtain a tear gas effect, a 'flash' of light."[69] In 1977, the Germans used a method similar to that proposed by Applegate to disarm hijackers in Somalia.[70] During the same period the British Special Air Service ordered the manufacture by Royal Ordnance Enfield of the stun grenade G60, to support its "counterrevolutionary" operations. Also called a "flash-bang," this grenade contained a mix of mercury and magnesium, which produced a blinding flash of 300,000 candelas (cd)[71] upon explosion and a boom of 160 dB.[72]

### The Flash-Bang Grenade

In the 1990s, the United States accelerated research into grenades through various organizations, in order to restock the explosives that had been in use for thirty years. The goal was to "combine a number of different effects to target multiple human senses"—in other words, to produce a sensory saturation that temporarily disables the target.[73] The Edgewood Research, Development and Engineering Center, one of several army research centers, was charged with devising "flash-bang" and smoke grenades.[74] The National Institute of Justice financed Sandia National Laboratories' development of a weapon projecting a cloud of combustible dust, which has desirable qualities when set alight, according to the manufacturer in its 2002 final report:

> At the point of detonation, the resulting flash-bang effects could be terrifying to an adversary. The target would be confronted with an exceptionally bright fireball at least two meters in width that would appear to envelop him totally. The acoustical report (170 dB) would probably create intense pain in the adversary's ears. The shock wave of 2.5 to 3.0 psi would create more terror. And if the ballistic contains a chemical irritant, it would cause the adversary even greater disorientation and discomfort.[75]

For its part, the JNLWD sought to obtain a "clear-a-space device," a device that would enable the evacuation of a zone without having to go there physically.[76] In June 2004 the Marine Corps kicked off its Clear-a-Space Distract/Disorient Program, asking industry to furnish projects.[77] SARA worked on its Multi-Sensory Grenade (MSG), which combined "sound, light and odor [to] overwhelm four of the senses. Unlike existing non-lethal weapons, the MSG's design allows for easy reconfiguration so that the sensory subcomponents can be changed to adapt to new uses."[78]

In the end it turned out to be the incapacitating grenade M84, produced by Universal Propulsion Company, a subsidiary of Goodrich, that is now used by the U.S. Army. The army described it, in 2003 in its "sources sought" notice, as "a non-lethal (stun) diversionary hand grenade, which produces an intense flash (approximately 1 to 2.5 million candle-power peak) and bang (approximately 170 to 180 decibels at 1.5 meters (5.0 ft.)). The grenade will be used by tactical and non-tactical forces while performing missions of hostage rescue and capture of criminals, terrorists, and other adversaries."[79]

In 2004, the U.S. Department of Justice signed a contract with ALS Technologies to furnish the Federal Bureau of Prisons with various grenades emitting up to 185 dB.[80] That same year the NIJ, the Bureau of Alcohol, Tobacco, Firearms and Explosives, and the National Tactical Officers Association asked the company E-Labs to evaluate the efficacy and security of eight "noise-flash diversionary devices," grenades with combined effects, the acoustic amplitude of which is between 161 dB and 181 dB. The study avoided drawing conclusions and mentioned that the makers of these grenades, which were already on the market, furnished a warning that they "may cause physical injury or death."[81] The report noted that two of these grenades could set fire to objects or cushions—and that all either displaced objects, cushions, or pillows when they exploded.

In fact, various unfortunate incidents occurred in the use of flash-bang grenades, affecting not only the designated targets but also law enforcement agents. In 2003, a fifty-seven-year-old Harlem woman died of a heart attack after an incapacitating grenade landed in her apartment; the police had invaded her dwelling by mistake, and the City of New York was required to pay $1.6 million to her family. In 2004, a sergeant on mission in Baghdad lost his right hand and was wounded in the leg by the sudden explosion of two grenades.

A marine corporal lost a finger when a grenade exploded in his hand. Another flash-bang was set off accidentally in a car occupied by three FBI agents: the vehicle caught fire, and one of the agents became deaf in one ear and suffered from persistent insomnia and migraines. In July 2009, the government was required to pay $29,000 in damages to a prisoner from Wisconsin who suffered persistent tinnitus after a grenade was thrown into his cell.

In 2008 Sandia National Laboratories, working for the National Nuclear Security Administration (which is part of the Department of Energy), returned to its flash-bang MK141, a twenty-year-old device, and launched an "improved flash-bang grenade" (IFBG) based on a project initially financed by the NIJ but never deployed.[82] The IFBG, conceived for maintaining order, was marketed by Defense Technologies. Contrary to prior flash-bangs, which used a mix of aluminum and potassium perchlorate, the IFBG uses a combustible-air explosive that produces only a limited shock wave while preserving the intensity of the flash and the bang (170 dB).[83] It seems the collateral damage issue was not entirely resolved, however: in 2010 the U.S. Coast Guard launched a new call for projects for a "non-pyrotechnic flash-bang grenade" in order to "diminish the risk associated with smoke and secondary fire,"[84] and in February 2011 a special weapons and tactics (SWAT) officer died in the explosion of an incapacitating grenade he was attempting to defuse at home.[85]

Russia, meanwhile, was not sitting idly by, and displayed several types of grenades during the Interpolitex weapons show of October 2005. The display of the Russian exporter Rosoboronexport was offering "the SV-1351 grenade, which produces combined psycho-physiological and mechanical effect on the perpetrators, with its intense flash, sound impulse and rubber shrapnel impact; the SV-1334 sound-and-flash hand grenade with automatic primer, that distracts and stuns criminals. . . . The inventory of the Russian special operations forces units also includes the Plamya-M sound-and-flash stationary grenades, Zarya-2, Fakel and Fakel-S sound-and-flash hand grenades."[86] The German company Rheinmetall Defence, which furnishes the British army, creates and markets not only offensive and defensive grenades that "generate a hyper-lethal blast wave" but also "1-, 2-, 6-, 7- and 9-bang versions," producing "near deafening sounds."[87] The list could go on, since a large number of countries (and private organizations) use stun grenades, from Israel to Albania, South Africa to Egypt, Kyrgyzstan to Colombia.[88]

### French Know-How

In France, "non-lethal" grenades, which are currently in use by law enforcement (police, army, gendarmes), are produced by the Etienne Lacroix Group, which in 1997 and 2006 bought Ruggieri and Alsetex, companies that until that point had furnished the state. The group covers pyrotechnics in its entirety: Lacroix-Ruggieri handles fireworks displays and is "the undisputed European leader" in pyrotechnics, while the Lacroix defense and security branch, which includes the weapons and research company Alsetex, handles materials used for repression.[89] Lacroix mentions in particular in its catalogue "less-lethal hand grenades" such as "sound grenades" and "multi-effect grenades," as well as the Cougar and Chouka grenade launchers, which can be used by law enforcement to launch grenades up to 200 meters.[90]

Alsatex's former website was more explicit in 2005, before the company was purchased: there were technical specs for the "teargas grenade GLI F4" (165 dB at 5 meters), the "GM2 flash grenade" (155 dB at 5 meters), the "stunning grenade SAE 420" (155 dB at 5 meters and 2 million cd for a "blinding effect that lasts twenty seconds"), the "deafening grenade SAE 430" (159 dB to 160 dB at 15 meters), or the "offensive grenade without shrapnel 410" (160 dB at 15 meters, which produces, according to the manufacturer, an "intense and psychologically aggressive effect" that "makes rapid and effective neutralization of protesters possible . . . in a context that is hard and resistant").[91] The French company Davey Bickford, for its part, presented grenades reaching 170 dB at 1 meter at the 2007 Milipol exhibition.[92]

In the Alsetex catalog there is also a ballistic dispersion device, otherwise called an "explosive grenade," a "disencirclement grenade," or a "manual protection device" (*dispositif manuel de protection* or DMP), the last of these names being used in official terminology.[93] Behind this abundant vocabulary is a weapon whose explosion disperses eighteen blocks of rubber (which allows one to "touch protesters in a circular manner, and to break the encirclement of the thrower") and which reaches a sound intensity of 145 dB according to Alsetex—or 165 dB according to another manufacturer, the Société d'application des procédés Lefebvre.[94] The DMP is described as having the effect of a "strong punch" or a "slap."[95] In January 2004, then interior minister Nicolas Sarkozy announced the introduction of the DMP at the same time as that of the Taser.[96]

Though they may be "non-lethal," these grenades certainly can mutilate thanks to shrapnel, shock waves, and, to a lesser degree, acoustic amplitude. Wounds due to the DMP include: a protester against nanotechnologies in Grenoble in 2006 who suffered a slashed cheek; the loss of an eye and of taste and smell for a young woman who was observing a protest in that same city in 2007; the amputation of two toes of a protester in Saint-Nazaire in 2009;[97] and, that same year, burns and wounds at the NATO countersummit in Strasbourg.[98] A few months later, the Commission nationale de déontologie de la sécurité (CNDS, National Commission on the Ethics of Security) published a study on the use of constraint and defense material by law enforcement in which it noted the numerous cuts and bruises brought about by the DMP during a 2008 protest in Grenoble.[99] It mentioned the directions for use ("launch by rolling on the ground") and cited a note by the director of public security dated December 24, 2004, which stated that the DMP should be employed only "in a framework of close self-defense and not to control a crowd from a distance." Additional wounds have been catalogued since, notably during a protest in Lorient (bruise to the eye when an attempt was made to "control a crowd from a distance") and at the gendarme training center at Saint-Astier (hearing problems).[100]

"Flash," "bang," "boom"—the effectiveness of explosives is embodied in these comic-book-style interjections. Brief in duration and often leaving no visible trace, such explosives nonetheless create more damage than the mythical infrasounds. The most famous weapons—the vortex or the plasma weapons—are also the most inoffensive, given that they have barely passed the experimental stage for the moment, or have proven to be poorly suited to the tasks for which they were intended. On the other hand, weapons that are still barely known, such as the shock wave cannon or DMP disencirclement grenades, seem to have a fine future ahead of them in the area of repression.

Anti-noise regulations often advise not surpassing 120 dB; independent experts situate the threshold for auditory pain at

140 dB and argue that a pulsing noise is potentially far more dangerous than a continuous noise. Weapons of medium or high frequencies, as we shall see, become controversial if they surpass 150 dB. And yet the manufacturers of "non-lethal" explosive weapons and their government customers boast of achieving amplitudes of 185 dB, with no ensuing debate. Similarly, while there exists an international protocol forbidding the use of blinding laser weapons,[101] deafening weapons and the "collateral damage" they produce are the subject of no specific legislation, and seem to ignore the international convention requiring proportionality (relative to the threat) and discrimination (between combatants and non-combatants) in the use of weapons.[102]

# 4

## "TOTALLY CUT OFF FROM THE KNOWN": SILENCE AND SATURATION[1]

When we go up a notch in the spectrum of frequencies, when we leave the strict territory of basses and shock waves to arrive at the domain of signifying sound (language, music), an entirely different story begins: it is no longer a matter of retracing the military-industrial genealogy of weapons systems but of delving into the "war of the mind," which plays out starting in World War II and continues today in other forms.[2] Sound is no longer employed solely for its essentially organic effects, but also used for its psychological impact; the distinction between the two, as we shall see, is not entirely clearly defined. Militaries and information services progressively institute what will become sensory deprivation techniques and develop new competencies to achieve this; psychologists, psychiatrists, neurologists, and other "mind doctors" become involved in war and police operations. Sound—or in some cases its carefully calibrated absence—becomes a weapon of submission, torture, and destruction.

Sensory deprivation can function either by neutralizing the senses or by bombarding them. Whether pitch blackness or full sunlight, antiseptic or stench-filled, total silence or constant sound, the desired effect is the same: to deprive a person of the use of his or her senses. Manipulation of the sound environment is part of a broader picture, neither more important nor more terrifying than other practices used simultaneously. Subjecting a person to sensory deprivation implies in particular that the person is placed in isolation, in other words, kept in a cell in which he has no contact with the outside world, other than prison or military personnel. In 1988, Nathalie Menigon, who was incarcerated at the Fleur-Merogis prison for belonging to an armed revolutionary group, Action directe, spoke about it as a "modern dungeon": "You find yourself in a void which, inevitably, gets inside of you."[3]

As scientists and the military launched the first broad research into mental control, they developed an interest in sensory deprivation in the form of silence. In the mid-1970s, a former American diplomat, John Marks, invoked the Freedom of Information Act to gain access to CIA archives and was given sixteen thousand previously classified documents. In 1979, he published *The Search for the "Manchurian Candidate"*, a work that retraces the CIA's history of experimentation in mind control.[4] In it, we learn that in the late 1940s, the brand-new CIA developed an ambitious program to understand the mechanics of consciousness and to master "behavior modification" in order to keep from losing ground to the Soviets when it came to extracting confessions. A "genuine hysteria over Communist mind control" spread throughout the United States at the beginning of the Cold War.[5] Chinese and Soviet methods, which were viewed as "the brainwashing equivalent of the atomic bomb," would drive the research ordered by the CIA.[6] Beginning in 1956, as it turned out, China and the USSR used "classic police tactics" that were no less violent: isolation, sleep and food deprivation, long periods of standing, and extreme temperatures.[7]

In *A Question of Torture*, historian Alfred McCoy analyzes the evolution of the CIA's interrogation methods "from the Cold War to the War on Terror." He indicates that at the beginning of the 1940s, the CIA was convinced that the communists were using drugs, electroshocks, or hypnosis, and believed that it was forced "to assume a more aggressive role in the development of these techniques."[8] Over the span of ten years, it would devote several billion dollars to testing "the mechanism of mass persuasion and the effects of coercion on individual consciousness"[9] through Operations Blue Bird and Artichoke.[10] In 1953, all of the programs were regrouped into Project MKUltra, under the direction of Dr. Sidney Gottlieb and counterespionage specialist Richard Helms. By 1963, MKUltra would invest $25 million in nearly two hundred projects and subprojects, undertaken jointly with Great Britain and Canada, and led by 185 independent researchers at eighty institutions, including forty-four universities and twelve hospitals.[11] In short, it was a very large program that, after the failure of experiments with drugs, quickly reoriented toward behavioral research.

The CIA, the ONR, and professors in experimental psychology worked in close collaboration to cultivate this field of study and develop what would become "no-touch torture": psychological torture.[12] In a similar vein to what happened later with the development of "non-lethal" weapons, the goal here was to make torture not less violent but less fatal, less visible, more effective, legally acceptable, and media friendly. Marks adds: "The intelligence community, including the CIA, changed the face of the scientific community during the 1950's and early 1960's by its interest in such experiments. Nearly every scientist on the frontiers of brain research found men from the secret agencies looking over his shoulders, impinging on the research."[13]

## THE "ACOUSTIC NIGHT": SILENCE[14]

Sensory deprivation was quickly placed at the heart of U.S. doctrine in the domain of mental control. While later "interrogation" methods would give priority to the use of sound saturation, it was silence that first got the attention of scholars and their patrons.

### The CIA and Experimental Deafness

Contract X-38 allowed the research council for the Canadian defense department to finance the first Canadian studies on manipulation of the senses.[15] In an article entitled "Experimental Deafness," which appeared in 1954, a psychologist at McGill University in Montreal, Dr. Donald O. Hebb, and two of his colleagues write about an experiment performed with six students who were paid to become deaf for three days.[16] Their ear canals were "packed with cotton impregnated with petroleum jelly"; then they were left to their usual occupations while asked to "keep a diary recording anything out of the ordinary that they observed about themselves." The results differed widely by subject, who, in addition to the physical discomfort, reported experiencing feelings of "inferiority, irritability, and tendency to avoid others." The researchers conclude that "a sudden lowering of normal auditory input has shown clear evidence in this experiment . . . of a disturbance in behavior."

That same year, Hebb's students published "Effects of Decreased Variation of the Sensory Environment."[17] Starting with the hypothesis that "the maintenance of normal, intelligent, adaptive behavior probably requires a continually varied sensory input," they intended "to examine cognitive functioning during prolonged perceptual isolation." Twenty-two young men were paid "to lie on a comfortable bed in a lighted cubicle 24 hours a day" with all their senses blocked: they wore "translucent goggles which transmitted diffuse light but prevented pattern vision," "gloves and cardboard cuffs." They remained in a "partially sound-proof cubicle," with "a U-shaped foam-rubber pillow in which the subject kept his head while in the cubicle," while "the continuous hum provided by fans, air-conditioner, and the amplifier leading to earphones in the pillow produced fairly efficient masking noise."

The result: "The subjects tended to spend the earlier part of the experimental session in sleep." Then they grew bored and impatient: "They would sing, whistle, talk to themselves, tap the cuffs together, or explore the cubicle with them. . . . There seemed to be unusual emotional instability during the experimental period. When doing tests, for instance, the subjects would seem very pleased when they did well, and upset if they had difficulty." On a cognitive level, "the subjects reported that they were unable to concentrate on any topic for long while in the cubicle." They had "blank periods" and hallucinations: "The visual phenomena were actually quite similar to what have been described for mescal intoxication. . . . One subject could hear the people speaking in his visual hallucinations, and another repeatedly heard the playing of a music box. . . . They said it was as if there were two bodies side by side in the cubicle." The students "also reported feelings of confusion, headaches, a mild nausea, and fatigue; these conditions persisted in some cases for 24 hours after the session." Most of the volunteers left at the end of two to three days.[18]

In 1955–56, at the National Institute of Mental Health (NIMH), near Washington, D.C., Dr. John C. Lilly immersed two volunteers in water-filled tanks, their eyes covered and the sound level reduced to a minimum. After only a few hours, two people developed hallucinations as well as "reveries and fantasies of a highly personal and emotionally-charged nature."[19] Enthusiastic about the experiment, the CIA sought to use this tank as a means to interrogate reticent subjects, in order to break them down "to the point where their belief systems or personalities could be altered."[20] But Lilly made a point of not conducting any experiments that would involve anyone other than himself or his colleagues. Understanding that the CIA did not intend to use his research in a positive manner, and that he could no longer work as he pleased, he resigned from the NIMH in 1958.

In 1963, MKUltra was officially interrupted due to its mixed results and the ethical issues raised by experiments led on non-volunteer human subjects[21] by doctors or even agents with insufficient scientific competency.[22] But that same year, the CIA reported its conclusions in *KUBARK Counterintelligence Interrogation*, a manual that outlined the agency's interrogation methods.[23] Absolute silence along with "self inflicted pain" was one of the techniques of this new "no-touch" torture.[24] The manual indicated in particular that "the circumstances of detention are arranged to enhance within the subject his feelings of being cut off from the known and the reassuring, and of being plunged into the strange." It goes on: "The chief effect of arrest and detention, and particularly of solitary confinement, is to deprive the subject of many or most of the sights, sounds, tastes, smells, and tactile sensations to which he has grown accustomed."[25]

The definition of sensory deprivation is based, notably, on Lilly's analysis of the autobiographical stories of polar explorers and other solitary navigators. Quoting Lilly, the manual states that "the symptoms most commonly produced by

isolation are superstition, intense love of any other human being, perceiving inanimate objects as alive, hallucinations, and
delusions."[26] The manual concludes:

> The more completely the place of confinement eliminates
> sensory stimuli, the more rapidly and deeply will the inter-
> rogatee be affected. Results produced only after weeks or
> months of imprisonment in an ordinary cell can be dupli-
> cated in hours or days in a cell which has no light (or weak
> artificial light which never varies), which is sound-proofed,
> in which odors are eliminated, etc. An environment still
> more subject to control, such as a water-tank or iron-lung,
> is even more effective.[27]

The spread of these practices continued through several
channels, primary among them Project X and the Phoenix
Program: the former organized training in CIA counterinsurgency techniques for Central and South American soldiers
and torturers, and the latter instituted the use of these same
techniques in Vietnam in "terminal experiments" against
Vietcong prisoners.[28] In 1966, the agency sent an electro-
shock machine and three psychiatrists to the psychiatric hospital at Bien Hoa, north of Saigon, to test the technique of
"depatterning" developed by Dr. Donald Ewen Cameron (a
technique we shall describe later) in real conditions, on prisoners of war.

The teaching of "no-touch torture" occurred mainly via
institutions such as the army's School of the Americas, which
from 1966 to 1991 trained officers from Central and South
America in military interrogation, and the Office of Public
Safety, which from 1966 to 1974 taught CIA techniques to
police from forty-seven countries, including Brazil, South

Vietnam, Uruguay, Iran, and the Philippines.[29] Ultimately, the agency published seven new manuals, including the 1983 "Honduran Manual," officially called the *Human Resource Exploitation Training Manual*.[30] Domestically, the United States applies techniques from these manuals in high-security, twenty-first-century prisons, where prisoners on the highest watch are placed "in total isolation in soundproof cells with white walls."[31]

### The Silent German Section

In the Federal Republic of Germany, the study of sensory deprivation began in 1971. Inspired by the work of Hebb and his students, experiments were performed "in the psychiatry and neurology clinic of Hamburg-Eppendorf [university hospital], under the direction of the Czech psychiatrist Jan Gross, who in the 1960s did similar experiments with Svab [a professor at the University of Hamburg]." The research program was cleverly labeled as an investigation of "psychosomatic, psychodiagnostic and therapeutic aspects of aggressivity" and was based on the "observation of human behaviour in situations of sensory deprivation."[32] A "silent chamber" was built within the university hospital for volunteers, recruited from among students and soldiers of the Bundeswehr. Reactions observed included fear and panic, distorted perception (hallucinations, autoscopy,[33] illusory manipulations), extreme hunger, troubled sleep patterns, chest pains, motor imbalances, shaking, and convulsions.[34]

Starting in October 1970, the government established a special detention system for prisoners of the Red Army Faction (Rote Armee Fraktion or RAF)—also known as the Baader-Meinhof Gang—and other extreme left-wing activists.[35] In 1972, exceptional treatment was instituted for

Astrid Proll and Gudrun Ensslin, and an especially lengthy treatment for Ulrike Meinhof, considered to be the group's ideologue.[36] They were incarcerated separately in a "special part of the silent section" of the prison at Cologne-Ossendorf:

> This wing is situated in one of the sections of the women's psychiatric building of the prison, separated from the main building and specially outfitted to be acoustically isolated. . . . The cells situated above and on the sides of theirs were left empty during their detention, so that no external noise could reach them. The walls and the furnishings of the cell were painted white and daylight penetrated into the cells only through a narrow slit covered with fine mesh. The prisoners from the special wing of Cologne-Ossendorf therefore lived twenty-four hours a day in no discernible setting.[37]

Ulrike Meinhof writes:

> Feeling your skull on the verge of breaking into pieces . . . feeling you cannot speak
> Impossible to recall the meaning of words, except very vaguely
> The whistling sounds—s, ss, tz, sch—, unbearable torture . . .
> Feeling time and space irredeemably
> interwoven one with the other and feeling yourself waver, trapped
> in a labyrinth of distorted mirrors
> And then: the terrible euphoria of hearing something which
> differentiates day from the acoustic night.[38]

In 1973, the prison psychologist, Professor Jarmer, noted about Ulrike Meinhof that "the psychic burden imposed on the prisoner goes well beyond the normally inevitable level for solitary confinement."[39] Committees against torture via isolation protested outside, demanding in particular the elimination of Cologne-Ossendorf's silent section. After five months in this section, Astrid Proll was returned to a normal cell in order to be fit enough to stand trial; "she finally had to be transferred to a sanatorium in 1975, her condition no longer allowing her to survive in detention."[40] Between 1972 and 1975, the Baader-Meinhof prisoners would lead several hunger strikes, unto death for Holger Meins, in order to demand the cessation of "special treatment" (*Sonderbehandlung*). Ulrike Meinhof, for her part, would hang herself in her cell in May 1976, in "disturbing" conditions, according to the international commission of inquiry.[41]

In an article entitled "The Silent Section, the Harshest Form of Solitary Confinement," one of the committees against torture writes:

> The same means of extermination are put in place and foreseen against political prisoners in other countries. A silent section is now under construction on the seventh floor of the Duivendrecht prison in the Netherlands. In Sweden, journalist Jan Guillou . . . was subject to similar torture in a silent section of the Stockholm prison. In the Portuguese colonies, the political prisoners are enclosed in cells made of tanks and immersed in water in order to cut off all noise and sensory stimulation.[42]

The therapist Sjef Teuns, insisting on the "key function" of "acoustic isolation," concludes: "Used for months and years, sensory deprivation is the perfect murder for which no one—or everyone, except the victim—is responsible."[43]

## "A GIANT TAPE RECORDER": SATURATION[44]

Although for many years CIA manuals favored silence as a means of achieving sensory deprivation, the agency also explored the potential of sound saturation beginning in the 1950s. Such saturation was studied in the context of the same research programs and by the same scientists.

### The 1950s and 1960s: Dr. Cameron

In 1954, a member of the CIA produced a report in which he recalled a presentation made by Dr. Hebb at the annual convention of the American Psychological Association regarding a Canadian army experiment that aimed to "eliminate as much as possible all sensation": the students, hands and feet covered in thick gloves, were placed in a silent room in which they were read "childish rhymes." The subjects "tended to lose their sense of time" and become "very irritable and requested time after time to hear the simple childish rhymes." No one lasted more than a week.[45]

The CIA also placed its hopes in Dr. Donald Ewen Cameron, president of the American Psychiatric Association, who was then the head of Allan Memorial, the psychiatric section of McGill University Health Center. There he developed research on "the effects upon human behavior of the repetition of verbal signals," which, with the help of drugs, can achieve an effect "analogous to the breaking down of the individual under continuous interrogation."[46] The agency "not only had a doctor willing to perform terminal experiments in sensory deprivation, but one with his own source of subjects."[47] Between 1957 and 1963, one hundred patients residing at Allan Memorial for psychological problems became involuntary subjects under Subproject 68 of MKUltra, financed through the Society for the Investigation of Human Ecology.[48]

Cameron used his patients—mainly women—to test a method of "depatterning": a combination of induced coma, electroshocks, and forced listening to repetitive messages, at first negative ("You never stood up for yourself against your mother or father . . . they used to call you 'crying Madeleine'"), then positive ("You will then be free to be a wife and mother just like other women").[49] The cassettes were played in a loop, sixteen hours a day, through "a football helmet clamped to the head for up to twenty-one days") or through loudspeakers placed beneath the pillow of patients in the "sleep room."[50] "We made sure they heard it," said a colleague of Cameron's. Cameron's primary assistant, Leonard Rubinstein, was an electronics technician who devised "a giant tape recorder that could play eight loops for eight patients at the same time."[51] The effect on the involuntary subjects was disastrous: loss of memory and of all independence.[52]

Cameron was satisfied with his findings and indicated that his methods of strict sensory isolation were "much more disturbing" than the voluntary deprivation Hebb's students underwent.[53] But by 1963, the CIA had grown impatient with waiting for usable results, the Canadian Defense Research Board "would have no part in" his work, and Hebb called Cameron "criminally stupid." It was not until 1988, when lawsuits were brought by former subjects, that the American Psychiatric Association expressed "our deep regret that psychiatric patients became unwitting participants in those experiments." As for the Canadian Psychiatric Association, far from apologizing, it applauded Allan Memorial's contribution to science.[54]

Great Britain, which also collaborated with the United States on MKUltra, reproduced the Canadian tests to verify them. In 1959, the medical journal *The Lancet* published a report of an experiment conducted at Lancaster Moor Hospital, a psychiatric institution, with twenty volunteers paid to remain in a soundproof room, glasses blurring their vision, gloves on their hands. Fourteen left within the first forty-eight hours; all experienced states of agitation, sometimes reaching panic; and five dreamed of "drowning, suffocating, and killing people."[55] The Royal Society of Medicine was split over a 1962 press review of past experiments—it described new ones, notably on schizophrenics placed in "a soundproof box belonging to the BBC," which "schizophrenics tolerated . . . remarkably well . . . their hallucinations were less vivid."[56] But as McCoy indicates: "Despite the surface legitimacy of its publication in *The Lancet*, the timing and cost of the experiment make it likely that this was military, not medical, research."[57]

### The 1970s and 1980s: Britain's "Five Techniques"

During the same period, the brutal methods of the British armed forces in colonies fighting for their independence were condemned, and in 1965 a "joint directive on military interrogation" was adopted, forbidding all use of violence and promoting "psychological attack." The military was officially trained in the new method, "offensively to conduct counterinsurgency, and defensively to survive the stress of capture."[58] In 1972, the Parker Report analyzed British interrogation procedures of individuals "suspected of terrorism" and indicated that these techniques were progressively refined following World War II "to deal with a number of situations involving internal security." It specifies:

Some or all have played an important part in counter insurgency operations in Palestine, Malaya, Kenya and Cyprus, and more recently in the British Cameroons (1960–61), Brunei (1963), British Guiana (1964), Aden (1964–67), Borneo/Malaysia (1965–66), the Persian Gulf (1970–71) and in Northern Ireland (1971).[59]

In December 1971, Ireland filed suit in the European Court of Human Rights for violation of the European Convention on Human Rights. Judgment in *Ireland v. the United Kingdom* would be rendered on January 18, 1978. With the number of bombings and deaths linked to conflict in Northern Ireland surging in the early 1970s, we find that to remedy the situation the English Intelligence Centre gave a seminar in April 1971 to the Royal Ulster Constabulary (the police of Northern Ireland) teaching new methods of "psychological attack," later known as the "five techniques." On August 9 that same year, Great Britain launched Operation Demetrius, a campaign of massive arrests of alleged members and sympathizers of the Irish Republic Army (IRA) lasting several months, and created a special system of "extrajudicial deprivation of liberty" for them.

Twelve people arrested on August 9, 1971, and two others detained in October would receive special treatment: shortly after their incarceration in "unidentified interrogation centers, these fourteen detainees would be subjected for a week to "a form of 'interrogation in depth' which involved the combined application of five particular techniques . . . sometimes termed 'disorientation' or 'sensory deprivation' techniques."[60] Other prisoners were threatened with the same treatment, this "interrogation in depth" serving both to break those who were subjected to it and to create a general climate of fear and intimidation.[61] The "five techniques," described by the European Court, were the following:

> (a) wall-standing: forcing the detainees to remain for periods of some hours in a "stress position" . . . ; (b) hooding: putting a black or navy colored bag over the detainees' heads . . . ; (c) subjection to noise: pending their interrogations, holding the detainees in a room where there was a continuous loud and hissing noise; (d) deprivation of sleep: pending their interrogations . . . ; (e) deprivation of food and drink.[62]

The European Commission would describe the third technique in a slightly different manner: subjecting the detainees to "continuous and monotonous noise . . . of a volume calculated to isolate them from communication."[63] Other testimonies describe a sound "like the escaping of compressed air or the constant whirring of a helicopter blade."[64]

The commission noted that "the combined application of methods which prevent the use of the senses, especially the eyes and the ears, directly affects the personality physically and mentally. . . . Those most firmly resistant might give in at an early stage when subjected to this sophisticated method to break or even eliminate the will."[65] It concluded that the use of these "five techniques" constitutes a case "in breach of Article 3 of the Convention in the form not only of inhuman and degrading treatment but also of torture,"[66] and it spoke of a "modern system of torture."[67] The European Court of Human Rights, arguing that there is a difference in intensity with respect to torture, also found a violation of the Human Rights Convention, but only for inhuman and degrading treatment.[68] Beginning in 1972, the British government agreed to stop using these techniques, whether individually or together. The fourteen detainees received between £10,000 and £25,000 each, as compared to a few hundred pounds for the detainees who were subjected to extrajudicial detention without application of the "five techniques."[69] A university psychiatrist judged that three of the detainees had become "psychotic" in twenty-four hours and were suffering from symptoms ranging from hallucination to "profound apprehension and depression."[70]

In France, while the practice of isolation is common to break recalcitrant detainees, there are no documented occurrences with regard to the use of sensory deprivation. Hellyette Bess, imprisoned in 1974 at Fleury-Merogis for her involvement with the group Action directe, had to listen to a "broken radio" for days, which the penitentiary administration said it couldn't turn off—but the episode remained an isolated one, according to Bess, an individual and "anecdotal" initiative.[71]

### The 1990s and 2000s: Israel and China

In 1998, a report by the Israeli NGO B'Tselem entitled "Routine Torture: Interrogation Methods of the General Security Service [GSS]" describes the interrogation techniques used on Palestinian prisoners by the Israeli armed forces, notably by the GSS.[72] The latter used a series of techniques called *shabeh*:[73]

> *Shabeh* is the combination of methods, used for prolonged periods, entailing sensory isolation, sleep deprivation, and inflicting pain. Regular *shabeh* entails shackling the interrogee's hands and legs to a small chair, angled to slant forward so that the interrogee cannot sit in a stable position. The interrogee's head is covered with an often filthy sack

and loud music is played non stop through loudspeakers.
Detainees in *shabeh* are not authorized to sleep.[74]

The state, according to its legal defenders, considered that
these methods are at worst "unpleasant."[75] They claimed that
playing music at high volume "is not done to oppress . . . but
to prevent interrogees from speaking with other detainees,"[76]
and added: "If there would be a budget to build a separate cell
for each [detainee], loud music would not be played."[77]

Nonetheless, the UN judged in 1997 that these procedures constituted torture, and demanded their cessation.[78] In
1999, the Israeli High Court of Justice called them "unac-
ceptable and prohibited," but it invoked the "necessity clause,"
which may absolve torturers in certain cases.[79] In fact, in the
2000s, none of the charges of torture or inhuman and degrading treatment led to any condemnation of the security
forces.[80] The practice therefore continued with variations
that enabled international law to be violated with impunity.
As far as sound goes, for example, music was replaced by anxi-
ety-producing sounds, mainly shouting.[81] A Palestinian testi-
fied in a 2007 report that "all the time there were noises in the
cell—knocking at the door . . . and I would even hear my own
screams during the interrogation, which they had apparently
taped."[82] In the Petah Tikvah prison, "at least one of the cells
is completely soundproof. In the other cells, the detainees
could hear disturbing sounds, such as monotonous dripping
of water on tin or the banging of metal doors."[83] This special
treatment was not applied to all the prisoners, but reserved for
Palestinians from the occupied territories."[84]

In China, a particular treatment was reserved for adher-
ents of Falun Gong,[85] who were subjected to "audiovisual
programs denouncing their movement" and forced "to listen

to student music full blast if they won't renounce their convictions."[86] According to a 2010 humanitarian report by the French branch of Action by Christians for the Abolition of Torture, "any Chinese citizen arrested or detained runs a significant risk of torture"—especially if he or she is involved in a non-patriotic military, professional, religious, or political activity, or is part of an ethnic minority.[87] Meanwhile, "the torture takes place in police stations, detention and investigation centers, prisons (*laogai* and *laojiao*) and secret detention centers.[88] It includes forced committal to psychiatric hospitals (*ankang*), the incidence of which has increased in recent years." Physical abuse was used along with techniques of sensory deprivation:

> The most commonly used methods are exposure to extreme temperatures, the obligation to remain several hours in painful positions, the prolonged use of handcuffs or ankle chains, exposure to violent noises or to blinding light, the privation of food and water, of sleep, of hygiene, of sensory stimulation (use of blindfolds or hoods, detention in a dark room), isolation for prolonged periods that can last several years.[89]

Moreover, while the law punishes the use of physical torture (*kuxing*)—to little effect, it seems—psychological torture is not recognized as such and therefore does not constitute an infraction of the law.[90]

### The United States: "Enhanced Interrogation"
In the 2000s, the United States put the earlier experiments in MKUltra to use, systematizing the use of sensory deprivation: "The White House made torture its secret weapon

in the war on terror."[91] In January 2002, John Yoo, an official at the Justice Department, circulated a memo affirming that the Geneva Conventions, which banned the use of torture, did not apply to the conflict in Afghanistan, which he labeled a "failed state," or to "illegal enemy combatants." In August of that year, in a second memo, Assistant Attorney General Jay Bybee redefined torture in a more restrictive sense—the sensory deprivation formalized by the CIA no longer, in his eyes, constituted an act of torture. In October, in a third memo, General James T. Hill recommended new techniques to deal with certain detainees at the U.S. military prison at Guantánamo Bay, Cuba, including "stress positions," isolation, "deprivation of light and auditory stimuli," hooding, use of twenty-hour interrogation, wet towels, and dripping water to induce a sense of suffocation. That same year, a Justice Department directive authorized the use of "waterboarding."[92]

In September 2003, in a fourth memo, Lieutenant General Ricardo Sanchez, the commander in Iraq, authorized twelve interrogation techniques supplemental to those described in the official army manual, *Field Manual 34-52*, notably sleep deprivation, deception, the use of military dogs to scare detainees, and "yelling, loud music and light control."[93] This memo would be considered so extreme that it would be revoked a month later—but it would leave its mark on military practices. "No-touch torture" under the euphemism of "enhanced interrogation techniques" was officially instituted in U.S. prisons in Afghanistan, in Iraq, at Guantánamo, and in other countries to which the CIA delegated some of its activities and detainees, such as Morocco and Egypt. In 2006, after a virulent public debate on the use of torture, the Pentagon again pulled back on the authorized techniques.[94]

The influence of the *KUBARK Counterintelligence Interrogation* and of the "Honduran Manual" is palpable in this rebirth of "no-touch torture"—but instead of utilizing total silence, sensory deprivation now relies on playing deafening music for the detainee, who is incarcerated in an extrajudicial framework and is usually "chained into a 'stress position,' in a pitch-black space made uncomfortably hot or cold."[95] In 2005, *New Yorker* journalist Jane Mayer noted the presence at Guantánamo and in secret CIA prisons of BSCTs (Behavioral Science Consultation Teams), scientific successors to similar groups of the 1960s. General Geoffrey Miller, tired of what he deemed to be excessive moderation in interrogations at Guantánamo, introduced the first teams starting in 2003. These psychologists and psychiatrists were there not to help the detainees but to advise the military on interrogation techniques. To do this, they drew on training they had received through the Survival, Evasion, Resistance and Escape (SERE) program.[96]

Instituted after the Korean War to prepare U.S. personnel for the risk of capture in a "non-democratic" country, SERE involves training to resist torture. But some see it as an apprenticeship, spreading the most brutal techniques used at Guantánamo and in other prisons, notably "waterboarding," sexual and religious humiliation, and "noise stress." A representative of SERE told Mayer, "Trainees often think that the interrogation portion of the program will be the most grueling, but in fact for many trainees the worst moment is when they are made to listen to taped loops of cacophonous sounds. One of the most stress-inducing tapes is a recording of babies crying inconsolably. Another is a Yoko Ono album."[97]

That is how the detainees at Guantánamo, their eyes, hands, mouths, noses, and eyes covered, would come, strangely, to resemble the students of the late Dr. Hebb, finding themselves bombarded with very loud music and the sounds of crying babies and meowing cats, among other exceptional treatment. This uninterrupted sound offensive in secret prisons could last days, weeks, even months at a time. Sometimes, notably in Morocco, the sound was played through a helmet that the detainee, who was handcuffed, was forced to wear. More often, it was played through loudspeakers in the cell, or, as in Mosul in Iraq, in a metal container transformed into an interrogation room, which soldiers renamed "the Disco."[98] At another military base, near Al Qaim, a sergeant reported that the "the interrogation sound system . . . was so good they used it for the Fourth of July celebrations."[99] It was the *Clockwork Orange* approach, as a former CIA advisor would term it.[100]

Binyamin Mohamed, an Ethiopian suspected of belonging to Al Qaeda, who was incarcerated and tortured before being freed without charges, described his experience in "cell 17" of a Kabul prison:

It was pitch black no lights on in the rooms for most of the time. . . . They hung me up. I was allowed a few hours of sleep on the second day, then hung up again, this time for two days. My legs had swollen. My wrists and hands had gone numb. . . . There was loud music, [Eminem's] "Slim Shady" and Dr. Dre for 20 days. . . . [Then] they changed the sounds to horrible ghost laughter and Halloween sounds. [At one point, I was] chained to the rails for a fortnight. . . . The CIA worked on people, including me, day and night. . . . Plenty lost their minds. I could hear people knocking their heads against the walls and the doors, screaming their heads off. . . . I call it brainwashing.[101]

The former detainee also mentions hearing sounds of "thunder, the sounds of planes taking off, cackling laughter and horror sounds."[102]

Other sounds used to break the "terrorists" include tapes of babies crying, the song "I Love You" from the children's television show *Barney*, music by Christina Aguilera, Britney Spears, Queen, Metallica, Drowning Pool, and Nine Inch Nails,[103] and more generally music that the prisoners, usually Muslims from Arab cultures, described as "unbearably loud," "infidel," or "Western."[104] Jonathan Pieslak, an American academic who works on the relationship between music and the soldiers in Iraq, was told by one of the soldiers, C.J. Grisham, that

> Metallica worked really well and really any kind of American music, except for the popular stuff. . . . You put on a Michael Jackson tape, which to me would make me talk, but you put it on those guys and like, "Oh, Michael Jackson," and it doesn't do anything for them. But you put on the hardcore, heavy metal American music from the Deep South or wherever. They don't want to hear that stuff, they think it's Satanic.[105]

The children's songs and the disco, pop, or sexually suggestive hits aim to harass, humiliate, or shock detainees. For Pieslak, who aims to preserve a position of "neutrality" on this use of music, it's not by chance that heavy metal figures frequently in this type of plan: "the distorted guitar sound and vocal articulation appear to be the most significant in causing the reaction of frustration and irritation."[106] And he quotes Sheila Whiteley, a British academic, who analyzed the physical specificities of the sounds produced by electric guitars in progressive rock: "Naturally produced sound waves have only a few harmonics, but these 'clipped' [distorted] waves have many, especially at a high level and this is what gives off the piercingly painful effect. Natural guitar sounds at loud volume are . . . far less aggressive."[107] For Pieslak, rap and heavy metal share the common characteristic of expressing power, aggressiveness, and violence. In "Music as Torture/Music as Weapon," the musicologist Suzanne Cusick, who takes a strong stance against torture and has contributed to bringing this use of music to light, speaks of a "struggle of masculinities" and talks about the creation of a sound field composed of "musics that those who don't identify with them often hear as embodying the sounds of masculine rage."[108]

In 2003, Sergeant Mark Hadsell explained that the objective was "to break a prisoner's resistance through sleep deprivation and playing music that was culturally offensive to them. These people haven't heard heavy metal. They can't take it. If you play it for twenty-four hours, your brain and body functions start to slide, your train of thought slows down, and your will is broken. That's when we come in and talk to them."[109] As a former detainee at Guantánamo, Ruhal Ahmed, explains, the sound fills the spirit; it prevents a person from thinking freely, from pulling away and recuperating from other forms of torture:

> I can bear being beaten up, it's not a problem. Once you accept that you're going to go into the interrogation room and be beaten up, it's fine. You can prepare yourself mentally. But when you're being psychologically tortured, you can't. . . . You lose the plot, and it's very scary to think that you might go crazy because of all the music, because of the

loud noise, and because after a while you don't hear the lyrics at all, all you hear is heavy banging.[110]

Donald Vance, a U.S. citizen imprisoned at Camp Cropper in
Baghdad, says the experience "sort of removes you from you.
You can no longer formulate your own thoughts when you're
in an environment like that."[111]

The NGO Reprieve, which defends prisoners "from
death row to Guantánamo Bay," has abundantly documented
the use of music in the "war on terror." For Zero dB, a "silent protest" against the use of music as a means of torture,
it has enlisted the musicians whose pieces were used as part
of this torture.[112] In July 2009, Zero dB demanded an end to
the practice in a letter to President Barack Obama.[113] Two
years later, no official prohibition had yet been formulated.
Though it was alleged "that DoD [Department of Defense]
interrogators improperly played loud music and yelled loudly
at detainees," it was determined that these techniques were
authorized in *Field Manual 34-52* under the categories of "in-
centive" (granting a detainee request in exchange for information) and "futility" (convincing the detainee that resistance
to questioning is futile).[114] These techniques still figure in the
revision of the manual released after the revelations of abuse,
*FM 2-22.3 (FM 34-52).*[115]

As for the participation of psychiatrists and psychologists
in the torture: as with Project MKUltra, such participation
would get mixed reactions from the medical community. The
NGO Physicians for Human Rights, surveying the question of psychological torture,[116] published a white paper in
2010, "Human Subject Research and Experimentation in the
'Enhanced' Interrogation Program," and denounced the role
of health professionals in the development of more efficient

and legally irreproachable torture methods.[117] The American Medical Association advised its members to refrain from these practices, reminding them of the ethics of their profession. The American Psychiatric Association forbade its affiliates from having any "direct participation" in interrogation.[118] Meanwhile, the American Psychological Association, in recognition of the rights of its members to take part in "national security endeavors," maintains, in McCoy's words, "stricter, more specific standards for the treatment of laboratory animals than for human subjects."[119]

Established during the fight against communism, the technique of sensory deprivation is today part of the arsenal deployed by the United States in the "war on terror," by China against political and cultural dissidents, and by Israel against Palestinians. An exceptional treatment reserved for opponents considered the most dangerous or most troublesome, sensory deprivation has an explicitly political character from the start: it selects and destroys. By enlisting "mind doctors" in military or law enforcement operations, this treatment profoundly influences the development of new weapons, notably those termed "non-lethal," and the emergence of a behavioralist conception of law enforcement: it is no longer a matter of killing, but one of "modifying behavior." Power no longer is applied solely to the outside but aims to punish the inside, to subjugate the spirit along with the body. And, as it spreads, power is also less visible: the use of silence or music leaves no visible traces, and aims, as we shall see, to minimize public perception of torture and make it acceptable.

# 5

## "HELL'S BELLS":
## MEDIUM-FREQUENCY SOUNDS[1]

The music used in the limited framework of "enhanced interrogation" was also used in the broader theater of military operations, mainly by the United States, beginning in the late twentieth century. In truth, while sensory deprivation is a recent invention, music and language did not await modern times to be exploited in a military framework. The first reference is even biblical, as the walls of Jericho were knocked down by the Israelites through cries and ritual trumpets: "So the people shouted when the priests blew with the trumpets: and it came to pass, when the people heard the sound of the trumpet, and the people shouted with a great shout, that the wall fell down flat, so that the people went up into the city, every man straight before him, and they took the city."[2] As for the Celts in the iron age, they used carnyxes, long bronze pipes ending in a little horn in the shape of an animal's head: the players marched at the head of military formations, and "according to written accounts of the time, some Roman soldiers ran in terror from the horrible sound."[3] In the nineteenth century, their Scottish descendants in the Great Highland Bagpipes would be officially integrated into the British army to better rally the troops and frighten the enemy.

An old Chinese tale tells how soldiers attached hollowed-out reeds to kites and flew them at night to exasperate the enemy with the wailing or moaning sounds they produced.[4] Sound can even make a good hunting instrument. In his *Petite histoire de l'acoustique*, Pierre Liénard cites a study on Chinese music therapy and comments jokingly, "In Asia, they know how to kill a bird from a distance with the sound of a cry: 'Sound must reach the bird's auditory nerve just at the quarter second that separates inspiration and expiration. . . .' Which presupposes that one is calculating the speed of sound, and the bird's Mach speed."[5] In the twentieth century, Liénard reports, "the acoustic experts at the National Institute of Agronomic Research (INRA) had . . . the idea of recording and broadcasting on loudspeakers the distress cries of the crow . . . which, played in a field, immediately prompted their fellow crows to fly away."[6] This technique has since been used with some variation by the makers of "scare cannons," who use the sounds of birds flapping their wings, the sound of a wounded bird, and the cries of predators.[7]

The birds may have a comeback: Roman Vinokur, in "Acoustic Noise as a Non-Lethal Weapon," points to the Russian epic poem *Bylina*, and in particular to the meeting of Ilya Muromets, the hero, and Nightingale, the thief, "a fantastic creature, half bird of the forest, half man of the steppe": "Ilya of Murom rode off through the forests of Bryn. Nightingale heard the thud of the hero's horse and whistled loudly. The horse stumbled under Muromets. Ilya said to his good horse: 'Have you never ridden through dark forests, have you never heard birds whistle?'" After this lesson in bravery, Ilya kills Nightingale with an arrow to the eye. Vinokur continues, "This fable goes back to the twelfth century. However, actual robbers, found frequently in the Russian forests up to the eighteenth century, commonly utilized a stun whistle to intimidate traveling merchants and their servants. The psychological effect was so significant that even the armed travelers preferred to give up with no resistance."[8]

In Jamaica at the end of the eighteenth century, villagers communicated by means of the abeng, a cow's horn "reproducing the pitch and rhythmic patterns of a fairly small number of Twi words, from their mother language."[9] The black maroons, escaped slaves who built independent communities, used it during guerrilla attacks on the colonizers "to scare the British with its 'hideous and terrible' dislocated tones, sometimes managing to repel the invaders with sound itself. Gradually, as the British learned to assign a cause to its shrieking, high-pitched sound, their terror of Maroon ambush only intensified."[10] In the first half of the nineteenth century, German physicist Thomas Johann Seebeck invented the siren, which the composer Murray Schafer calls "a centrifugal sound, designed to scatter people in its path."[11]

Drums, trumpets, military chants, goose stepping by hundreds of soldiers, helmeted and booted policemen simultaneously tapping their clubs on their shields before a charge: the same sounds, the same rhythms manage both to avert and intimidate the enemy and to galvanize one's own troops. Sound offers both a demonstration of power and a coded form of communication. Long after the picturesque anecdotes just mentioned—amusing, perhaps, due to their distance in time, and no doubt as well because in that era adversaries were still fighting, acoustically speaking, with equally matched weapons—we witness, beginning with World War II, the development of the use of sound as an instrument of domination, then, as technology and political factors evolved and as sensory deprivation methods were developed, as an instrument of control.

Medium and high frequencies, music and sirens are treated together here, since they share the same military and law enforcement history, and the same devices are used to broadcast them. That said, the history of their usage follows two distinct threads: on one hand, the broadcasting of sound via loudspeakers, essentially in wartime operations, as means of ensnarement or harassment; on the other, the quest for "repulsive sounds," sounds that are immediately unbearable. The two uses ultimately come together in the construction of an efficient sound space, composed of "ghost sounds" that are at times audible, inaudible, unbearable, or ensnaring, depending on their designated targets and objectives. Contrary to weapons using low frequencies, the weapons used here in an offensive manner are often not developed specifically or exclusively for this effect. Their use falls along various points on the spectrum, by turns civil, military, utilitarian, diversionary, coercive, or "pre-lethal."

## "NOWHERE TO RUN": COMBAT LOUDSPEAKERS

Beginning with World War II, two evolutions occurred: first the recurrent use of loudspeakers as part of the military operations of both sides, then the emergence of what would soon be called, in the United States, psychological operations, or psyops, which would progressively transform what had been a mere propaganda machine into a complex system of information and disinformation, with a specific command structure, units, and weapons.[12] According to the U.S. system of categorization, loudspeakers are classified in the same category as printed tracts or "face-to-face" approaches, as instruments of "tactical psychological operations" with local reach.[13] Voice, music, and sound combine, at times to transmit instructions, at times for their physical properties, often to affect the body and mind simultaneously. The theory of "non-lethality," which would be formulated as such only in the 1990s, shows its first stirrings here. The military loudspeakers were also, as we shall see, the precursors to modern stereos—and were the manifestation of a growing interaction, in the second half of the twentieth century, between the military-industrial complex and the entertainment industry.

### World War II

In August 1941, in Ukraine, a long column of refugees and soldiers from the Red Army was fleeing toward the east. Suddenly, "German J-87 dive bombers (Stukas) appeared in the sky and began bombing and strafing the road, with their sirens loudly screaming for a greater psychological effect. The attacked people fell to the ground or ran amuck in panic."[14] Marie-Catherine Villatoux and Paul Villatoux, two of the rare historians who have studied the history of military loudspeakers, indicate in *Voices from Heaven* that, starting with the "phony war" of the years 1939–40, special units of the Wehrmacht used loudspeakers on the ground as weapons of propaganda to demoralize enemy troops.[15] The technique was tested by the U.S. Army in North Africa and in the Pacific in 1944, when for the first time loudspeakers were placed on planes. At the time, the experiment left much to be desired: the messages were drowned out by the sound of the motors. The Japanese also broadcast sounds through loudspeakers, mainly to rattle the British.[16]

In the final years of the war, the Allies also put in place various diversionary tactics in order to fool the Germans as to the places and dates of the Normandy landing. In early 1944, as part of Operation Fortitude, inflatable tanks, a wooden infrastructure, radio scripts spoken by actors, and the spread of rumors gave the impression that General George S. Patton was getting ready to launch an attack from southern England. A few months later, another less well-known "ghost army" took action between Normandy and the Rhine: the U.S. 23rd Special Troops, which included artists, actors, sound engineers, graphic designers, photographers, and painters, recruited from New York art schools and Hollywood studios, created visual and sound effects to make the presence of divisions believable.[17]

Initiating this gigantic camouflage effort were Douglas Fairbanks Jr., the Hollywood actor and producer, and Hilton Howell Railey, a journalist and public relations expert. According to philosophy professor and sound theoretician Christoph Cox in "Edison's Warriors," "it was Fairbanks who was responsible for marketing 'sonic deception' to America's military brass": "In the early 1940s, a family friend, Lord Louis Mountbatten [then a British naval officer], told Fairbanks about a secret British unit, based in a castle in Scotland, which was experimenting with sound effects, broadcasting recordings of tanks, aircraft, armored cars, and soldiers' voices under the cover of fog or smoke screens." Mountbatten also told him that "the British had already experimented with sonic tactics in the North African desert, hiring an Egyptian film company to broadcast sound effects in an effort to confuse the Italians and Germans."[18]

In 1942, Fairbanks convinced the U.S. Navy of the utility of "sonic warfare," and recordings were prepared at Pine Camp, in upstate New York, before being deployed in the European theater. To train the troops in sonic deception, Fairbanks enlisted the help of sound engineer Harold Burris-Meyer. Burris-Meyer was then an adviser to the Muzak company[19]—which wanted to develop a factory sound environment conducive to increasing the productivity of workers—

and was also the recent inventor of the stereo sound system that would be used in Disney's *Fantasia*.[20] Starting in 1938, he had been interested in creating "mass hysteria" among spectators by means of sound.[21] Under the leadership of the National Defense Research Council (NDRC), Fairbanks and Burris-Meyer launched a project called "The Physiological and Psychological Effects of Sound on Men in Warfare."[22] Noting that the "screaming whine caused by a siren" of bombers in nosedives provoked "paralyzing panic" among the soldiers on the ground, the NDRC decided first to make a "sonic bomb." But when the idea failed to take form "they shifted their work toward battlefield deception" with the help of Harvey Fletcher, director of acoustic research at Bell Labs and the inventor, in the 1920s, of stereo headphones.[23]

All "military activity that could possibly be useful in mounting a deception" was recorded[24]—even the barking of dogs, after the engineer in charge of recording learned of a Japanese superstition associating this sound with imminent death.[25] Great care was taken in the recording, and Bell Labs furnished not only the equipment but also instructions for broadcasting according to the geography of the site and the weather.[26] The recordings were then stocked by the Army Experimental Station, directed by Railey. On the ground, "sonic cars" were positioned and moved according to the virtual battle plan to arrive at a "a perfect blend of surround-sound verisimilitude."[27] After the war, Railey wanted to continue his work on sound deception, but the army lost interest in the tactic; it cut funding, closed the Army Experimental Station in 1945, and forbade soldiers to speak about the operation for at least fifty years. Not until 1988 did the army produce a field manual recommending the revitalization of this "lost art."[28]

The effort involved in perfecting these virtual sounds would have a far greater impact than that of temporarily confusing the Germans: Bell Labs would later note that "the new military acoustic devices were not just copies or minor physical modifications of existing instruments . . . but rather basically new designs," the acoustic fidelity of which would "become popular in the civilian world after the war in stereo hi-fi systems and studio monitors."[29] As early as 1942, Burris-Meyer predicted that the military sound techniques would be used in cinemas after the war. He contributed to the Defense Department's development of better-quality airborne loudspeakers[30] and attempted as well, apparently without success, to develop a sound torpedo: "Fired from a submarine, the torpedo would travel a distance and then surface. When the timer hit zero, the torpedo would eject the speaker and start the tape recorder, which would play a program of sound effects and then self-destruct."[31]

### The 1950s and the Vietnam War

A veteran of the U.S. Navy who participated in the first aerial operations with loudspeakers in 1944 relaunched their use during the Korean War in 1950: a plane called *The Voice of the United Nations*, decorated with a bear standing in front of a gramophone, had its first test flight, and it was soon joined by *The Speaker*. But the messages broadcast were hardly audible, and the army again abandoned this form of propaganda in 1952.[32] That same year, having caught wind of the American experiments, the British performed tests in Malaysia, basically broadcasting slogans in a loop; for the first time, a tape replaced live operators. As for the French, at the end of the 1940s and then in 1950, they tried loudspeakers on the ground and in the air, alternating messages, music, and the distribution of pamphlets to maintain order in overseas territories and colonies. They did this first in Indochina, then in Algeria, with tactical "revolutionary war" units using loudspeakers and distributing tracts. But once again, the experiment left much to be desired.[33]

The true launch of aerial propaganda took place during the Vietnam War. The film world recorded this spectacular deployment in *Apocalypse Now*, directed by Francis Ford Coppola, in which a full-blast "Ride of the Valkyries" announces the presence of the U.S. Army in the skies over Vietnam.[34] C-47 airplanes were systematically equipped with loudspeakers in the context of the psychological programs Chieu Hoi (which aimed to encourage defectors) and Quick Speak. A new system of broadcasting was used for the occasion: the HPS-1, composed of an amplifier and four powerful speakers, capable of projecting sound up to 4 kilometers. According to Colonel Rex Applegate, the HPS-1 was also used by the Canadian Ministry of Defense, the Iranian and Ethiopian air forces, and the Madrid, New York, and San Diego police.[35]

Another novelty during the Vietnam War was that planes were sent out on nocturnal "harassment" missions, and beginning in 1967, helicopters were also equipped with loudspeakers.[36] One of the best-known cassettes among those broadcast during these missions was "The Wandering Soul,"[37] a mix of "ghostly sounds; children crying because their father's body was not buried; ancestors calling out for soldiers to surrender or desert; Buddhist funeral music."[38] The voices were recorded in echo chambers to give the sense of coming from the beyond, while the roaring of a tiger was recorded at the Bangkok zoo.[39]

The helicopters that broadcast these recordings generally were shot at quite a bit, not out of superstition but because the sounds, played by night and at high volume, audible everywhere and continuously, became unbearable: "As one Viet Cong commander complained, these audio messages were hard to ignore, for the sound even penetrated through the earth to VC hidden in underground tunnels."[40] And Bill Rutledge, a pilot engaged in these operations, concluded: "Killing was our business and the PSYOP tapes helped make business damn good."[41] The account of an officer referring to another tape offered a similar perspective: "He recorded Robert Brown's 'Fire' from 1968 and used the 'demonic' portion repeatedly in an endless loop. He mentioned that the tapes often enraged the Viet Cong and led directly to their death."[42]

### The 1980s and 1990s

Helicopters equipped with loudspeakers continued to be used in the 1980s and 1990s in the course of humanitarian-military operations in Grenada, Somalia, Haiti, and Bosnia-Herzegovina.[43] General Tommy Franks relates that during Operation Desert Storm, in Iraq in 1991, "Every night, psychological operations units drove trucks fitted with gigantic loudspeakers slowly back and forth along the border, playing recordings of clanking tanks and Bradleys."[44] Other testimony speaks of "a little known deception operation that was carried out by a military organization clandestinely known as Task Force Troy. . . . Task Force Troy was given cover responsibility for an area of the Kuwaiti front which would normally have been covered by a full division. . . . The unit relied on the use of deceptive decoys, such as armored vehicles, artillery pieces and helicopters."[45]

In 1989–90, Operation Just Cause culminated in the invasion of Panama by the United States, in an effort to rid the country of General Manuel Noriega, then in power. By late December 1989 Noriega had taken refuge at the Vatican embassy, where General Carl Stiner, who was in charge of U.S. operations, "erected a sound barrier" that consisted of bombarding the building—and the surrounding neighborhood—day and night with hard rock and heavy metal, notably AC/DC, Mötley Crüe, Led Zeppelin, and Metallica, and "Nowhere to Run" by Martha Reeves and the Vandellas.[46] A soldier would later explain that the goal was to "limit [Noriega's] communications."[47] But the music also played another role, as Ben Abel, spokesman for the Psychological Operations Command at Fort Bragg, explains that Noriega, an opera lover, "commented that the rock 'n' roll was bothering him. Once the guys found that out, they cranked it up even more."[48] The official, more cautious version suggests that the idea was to prevent reporters from using "powerful microphones to eavesdrop on delicate negotiations."[49] But criticism from the Vatican and from Catholics came pouring in, and the noise was stopped on orders from General Colin Powell and President George H.W. Bush, who found the actions "politically embarrassing," as well as "irritating and petty."[50] Abel confirms that "since the Noriega incident, you've been seeing an increased use of loudspeakers. The Army has invested a lot of money into getting speakers that are smaller and more durable, so the men can carry them on their backs."[51]

In 1993, in Waco, Texas, the members of the Branch Davidian sect led by David Koresh retreated to their ranch. After law enforcement operations failed, the FBI was called in to attempt to dislodge the group. Various techniques were used to this end, notably bright lighting, flash-bang grenades, and the broadcasting at full volume and for several days in a row of deafening sounds:

Annoying sounds such as dental drills, seagull squawks, shrieks of rabbits being slaughtered, sirens, telephone busy signals, crying babies, trains in tunnels, and low-flying helicopters, as well as jarring music including Tibetan Buddhist chants, reveille, marches, Mitch Miller Christmas carols, selections from Alice Cooper, and Nancy Sinatra's 1960s pop ode, "These Boots Were Made for Walking."[52]

Added to the mix were "tapes of previous negotiations, and messages from those who had exited the compound."[53] The FBI even considered calling in a Russian scientist who had allegedly developed "acoustic psycho-correction" techniques to distribute subliminal messages during telephone exchanges, but that idea was put aside.[54]

The strategic goal—which was no more effective than previous strategies, since it ended with the death of a number of Davidians in a fire—was to "deny the Branch Davidians creature comforts in an effort to secure their surrender."[55] To achieve this, the FBI called in a team of psychologists, psychiatrists, psycholinguists, behavioral scientists, and religion experts. The innovations of the MKUltra program, which had engaged "mind doctors" to develop sensory deprivation, were exploited, no longer to extract information by force from an individual but to persuade a group to turn itself over to the authorities. Regarding the FBI's "failure in the use of behavioral science," Alan Stone, a professor of psychiatry and law, commented in a report to the Department of Justice:

The constant stress overload is intended to lead to sleep-deprivation and psychological disorientation. In predisposed individuals the combination of physiological disruption and psychological stress can also lead to mood disturbances, transient hallucinations and paranoid ideation. If the constant noise exceeds 105 decibels, it can produce nerve deafness in children as well as in adults.[56]

### The 2000s and the War in Iraq

During the war in Iraq that began in 2003, music was used on a large scale by the United States as a means of torture, as we saw, but also in theaters of operation. GIs spoke of the role of MP3 players to help them relax or, on the contrary, prepare for battle: "I'm going to have to shoot at someone today, so might as well get pumped up for it."[57] They also mentioned the installation of sound systems in armored vehicles and listening to music to stay alert during missions.[58] Trucks with loudspeakers were used to broadcast calls for surrender, but also to lead "harassment operations": over a period of days and nights, hard rock, heavy metal, and rap were played full volume. The choice of this type of music was motivated, according to Pieslak, by two primary factors: the soldiers considered these types of music "to be immediately irritating to insurgents in Iraq," and they drew inspiration from the Noriega precedent.[59] Real-life homage was paid to Coppola during a raid in Baghdad during which "Ride of the Valkyries" was blasted outside armored vehicles, as in the movie.

During the siege of Fallujah in 2004, "Hell's Bells" by AC/DC and "Welcome to the Jungle" by Guns N' Roses were blasted through the streets of the city. "Not to be outdone, the mullahs responded with loudspeakers hooked to generators, trying to drown out Eminem with prayers, chants of *Allahu Akbar*, and Arabic music."[60] The Marines nicknamed the city, pejoratively, "Lala Fallujah," and U.S. Army spokesman Ben Abel noted:

These harassment missions work especially well in urban settings like Fallujah. . . . The sounds just keep reverberating off the walls. . . . It's not the music so much as the sound. It's like throwing a smoke bomb. The aim is to disorient and confuse the enemy to gain a tactical advantage. . . . If you can bother the enemy through the night, it degrades their ability to fight.[61]

These objectives and uses of sound hark back to the days of CIA interrogators using "no-touch torture." Suzanne Cusick nonetheless notes a distinction: "Theorists of battlefield use emphasize sound's bodily effects, while theorists of the interrogation room focus on the capacity of sound and music to destroy subjectivity."[62] The cultural content of the music allows for a violent affirmation of one's own identity while manifesting the will to crush that of the other. As for the deafening volume, it aims to give a feeling of all-powerfulness that is both technological and ethereal: the sound falls from the sky, and no one can escape from it. According to Dan Kuehl, a retired U.S. Air Force colonel and psychological operations trainer, "Almost anything you do that demonstrates your omnipotence or lack of fear helps break the enemy down."[63]

Loudspeakers were also used in Fallujah to foster deception: "One TPT [tactical psyops team] used its loudspeaker to broadcast armored vehicle sounds to draw insurgents into an ambush."[64] The task was facilitated by the use of a new piece of equipment: the long-range acoustic device (LRAD). The LRAD, which can direct sound at a distance up to 3,000 meters, has a dual function: as a megaphone (distributor of voice or music) and as a siren (distributor of pure sound). We will describe more precisely its use to create a defensive barrier of sound, but we should note that the LRAD, which is used today in law enforcement as well as military operations, was developed by the U.S. Navy in its quest to improve its counterpiracy arsenal. Ships began using it in 2003 to intercept and halt foreign embarkation, and thus preserve a zone of exclusion 500 meters around.[65] The American army and various police departments also have available other forms of sound equipment: for example, the company Power Sonix, the "sound of Homeland Security," provides lower-amplitude devices.[66] The "high-power acoustic beam weapon," earlier dreamed up by the prolific company SARA, found belated expression in other devices with long names: high-intensity directed acoustics or acoustic hailing devices.

These loudspeakers have thus partially responded to the 2005 call of the U.S. Army's Special Operations Command, which noted an urgent need to update equipment in use since the 1990s and to develop a new family of loudspeakers.[67] The research continued in another direction as well, with the Special Operations Command calling for a "disposable speaker system capable of being airdropped or scattered and remotely operated."[68] In 2008, psyops aimed to implement a "next generation loudspeaker system" capable of furnishing "high quality recorded audio, live dissemination, and acoustic deception capability," including versions for drones, "Scatterable Media Long Duration (SMLD) variants," and "Scatterable Media Short Durations (SMSD) variants."[69] The Burris-Meyer "sound torpedo" apparently was not such a fantasy.

## "SANITIZE DISSENT": SOUND AVERSION[70]

While the LRAD and other contemporary devices are descendants of military loudspeakers, and therefore classified among propaganda instruments, they are also the latest in a line of weapons aiming to use medium and high frequencies to overpower individuals. This double filiation is today at the origin of virulent technical-legal debates seeking to determine if the LRAD and other "sound cannons" are tools of communication (as those who advocate a deregulation of their use would have it) or weapons (as their opponents, but also some of their promoters, affirm). It is no longer a matter so much of harassing with constant noise, but one of identifying sounds capable of provoking immediate deterrence, even physiological danger. Psychoacoustics, which studies the relationship between the physical properties of sound and human auditory sensation, is put to the service of domination.

According to Jürgen Altmann, when it comes to audible sound "annoyance can occur already at levels far below bodily discomfort, in particular if the sounds are disliked and/ or continue for a long time," but its effects generally are not lasting.[71] In the middle and high frequencies, it's mainly the intensity of the sound that matters: the risks of hearing loss are slim below 140 dB, although even brief exposure at this level can produce permanent loss of hearing in certain people. Above 140 dB, temporary loss of hearing and difficulties in equilibrium have been noted, even with sound protection. Above 160 dB and in the high frequencies, tickling and a sensation of heat can appear in the nose, in the mouth, on the skin, or, in the case of ultrasounds, on the hair. As a point of reference, emergency sirens (ambulances, police vehicles) are generally in the range of 123 dB and below.[72]

### From the 1960s to the 2000s

At the end of the 1960s, the HPS-1 used during the Vietnam War was sometimes combined with an auxiliary system called the "Curdler" or the "People Repeller," capable of emitting frequencies between 500 Hz and 5,000 Hz. Colonel Applegate describes it eloquently:

> The Curdler unit, utilizing the full 350-watts power, emits a shrieking, shrill, blatting, pulsating, penetrating sound equal to 120 DBM at 30 feet. It will break up slogan shouting, chanting, singing, hand clapping, rhythmic noise beats and agitator control. . . . At close ranges, the dissonant sound is so piercing that it forces advancing would-be rioters to turn away, discard their weapons, banners, signs, etc., in order to free their hands to cover and protect their ears. This sound effect is magnified when it is used in confined areas such as in narrow city streets, etc., where the sound bounces off buildings and pavements. The Curdler effect is very eerie at night and it can induce panic in many instances.[73]

In 1973, during the riots in Northern Ireland, the British bought thirteen of these loudspeakers from the American company Applied Electro Mechanics but don't seem to have used them at that time.[74] Around the same period, the University of Birmingham "seriously considered using an 'alarmingly loud' bell or a (scarcely audible) ultrasonic device to make life unbearable for students occupying the administration offices," but in in the end decided to refrain.[75]

In the 1970s, the United States conducted a program called Disperse, which aimed to develop "nonpermanently damaging crowd control devices" by means of light and sound.[76] A report by Harry Diamond Laboratories, now integrated into ARL,[77] indicated in 1975 that "there exists . . . sufficient technical information to support at least an exploratory investigation of . . . aversive audible acoustic stimuli, infrasonic and

ultrasonic systems, and bright flashing and flickering light."[78]
It recommended, in particular, the "evaluation of irritating or
pain-inducing sounds, such as the sound of fingernails being
scraped down a chalkboard, or very loud sirens, as non-lethal
weapons."[79] The report did not lead to any concrete action: "it
is assumed that these experiments were not performed or did
not produce positive results."[80] But the concept of "aversive
audible acoustic stimuli" was forged along the way.

In the early 1990s, the United States was interested in
the development of "tunable" acoustic weapons that "would
offer incremental penalties (e.g., capable of inflicting discomfort, incapacitation or of having a lethal effect)."[81] In 1995,
the company SARA produced a report for the army in which
it announced the possibility of releasing "nerve irritation, tissue trauma and acoustic fever (rise in body temperature)" by
means of frequencies between 500 Hz and 2,500 Hz, as well
as "a moderate to lethal rise in body temperature, tissue burns
and dehydration" with frequencies between 5,000 and 30,000
Hz.[82] But the projects seem to have remained on the drawing
board. Altmann, for his part, believes that "total immersion in
an ultrasound field above 180 dB would be required to over-
heat a human body to death after more than 50 minutes."[83]
That hasn't stopped SARA from continuing to publicize, even
today, without further details, its research into "high power
acoustic devices of all shapes and sizes ranging from sources
the size of a pencil to the size of a compact car."[84]

In 2000, ARDEC presented its Aversive Audible Acoustic
Device (A3D), a new name for the Gayl blaster. The presentation mentioned sound in a vague range of frequencies (between 15 Hz and 20,000 Hz) that would induce in those who
heard it "a strong desire to avoid [it] because of dislike, repug-
nance."[85] Although the A3D was developed by an engineer

in the advanced weapons technology section of ARDEC, the center insisted that the device should be "consider[ed] as a tool and *not* as a weapon." Then, in 2002, the army research center moved toward adapting the A3D to use as a "multi-sensory deprivation land mine" that would make "hearing protection *not* effective" and would use both "pure tones with adjustable frequency, amplitude and modulation" and "complex use/computer generated waves."[86] Finally, starting in 2003, ARDEC commissioned SMBI to attempt to identify sounds that were unbearable to hear, capable of "disrupting targeting, balance, and high-order cognitive processes in both humans and animals." Three years later, the laboratory lamented that "while our data suggests that there are unconditional aversive properties of sound, truly aversive sound has been elusive."[87]

## The LRAD

Where SMBI failed, a commercial company succeeded: American Technology Corporation (ATC), which renamed itself LRAD Corp. in 2010 and which, that same year, joined the top-ten best-performing U.S. defense stocks quoted on the stock exchange.[88] Things had not started out so rosily: at the request of the Defense Department, which after September 11, 2001, had wanted to possess a portable acoustic weapon that would disorient hijackers without damaging a plane's fuselage, the company had initially offered a "directed stick radiator," otherwise known by the more fanciful name of "acoustic bazooka." The idea was the following: into "a tube made of a polymer composite, around a meter long and four centimeters in diameter," are inserted "a series of piezoelectric discs,[89] each of which acts like a small speaker. Sending an electrical signal to the first disc at the rear end of the tube makes it expand, sending a pressure wave—a sound pulse—

along the tube. The pulse soon reaches the second disc, which is 'fired' at precisely the right time so that the sound pulse it produces magnifies the pressure wave," and so on down the tube until the end.[90] The president of the company, Elwood Norris, indicated that the weapon in its final version would emit a frequency between 6,000 Hz and 10,000 Hz at an amplitude of 140 dB, and concluded: "You could virtually knock a cow on its back with this."[91] As for humans, the military sales manager, Del Kintner, affirmed: "If you stand in the beam for more than 10 or 12 seconds, you get sick. People turn as green as grass, and you can pulse it in such a way that their ears don't really recover."[92] But the miraculous weapon and its "sonic bullets" soon joined the long list of abandoned prototypes: the years passed and "no results of any such testing have been published, to date, in the scientific literature."[93]

Next the company invented its signature product, the LRAD, which combines on a concentrated surface multiple little piezoelectric loudspeakers emitting in phase, allowing it to send a sound at an amplitude of 121 dB at 1 meter in its "voice" mode, and 153 dB at 1 meter in its "warning" mode (137 dB for the least powerful model), while choosing very precisely the direction of the emission and a distance of 600 to 3,000 meters (the acoustic intensity diminishing with distance). Various sources can feed the LRAD: a microphone to send instructions to an entire neighborhood, a CD or MP3 player, or a warning-tone generator (which produces a pure sound that, depending on the model, reaches between 1,000 Hz and 2,500 Hz, a mid-range frequency or higher, but in any case far from the ultrasounds that are regularly attributed to it).[94] At a distance of less than 100 meters, the warning mode makes any action other than fleeing or seeking protection difficult: earplugs and anti-noise protection helmets are of limited effect, and the only solutions are to leave the area or place oneself behind the device (the LRAD operators are not unduly affected, since the sound travels only forward), to the side of the device (the emission angle is between 15 and 30 degrees), or at a distance that diminishes the impact.

According to Elwood Norris in 2004, the LRAD "will produce the equivalent of an instant migraine" and "knock [some people] on their knees."[95] And according to Altmann, "in the warning mode the LRAD produces sound pressure levels which are dangerous to hearing if unprotected target subjects are exposed longer than certain durations: a few seconds at 50 m distance, 1.5 minutes at 100 m. Below about 5 m any exposure can produce permanent hearing damage."[96] Given the real risk of deafness, the promoters of the weapon claim speciously that no one would voluntarily remain in the field of maximum emission—but they don't bother considering those who might remain there against their will, such as the elderly, the handicapped, the wounded, or those kept there by force.

In 2004, the New York police deployed the LRAD, without using it, during the Republican National Convention. In 2005, the *Seabourn Spirit*, a cruise ship, used it against Somali pirates. That same year, the Santa Ana police used it to remove occupants from a house, and it was used in the zones destroyed by Hurricane Katrina in New Orleans to communicate with survivors. In 2007, the Georgian police used it against opponents in Tbilisi. In February 2009, a Japanese whaler used it against ecology activists from Sea Shepherd, who now also possess the equipment. In August 2009, the Thai police used it in Bangkok to disperse workers protesting layoffs at the Triumph factory. In September 2009, it was used by the Pittsburgh police against G20 activists—as well as by those involved in the putsch in Honduras, who, during a siege of the Brazilian embassy where President Manuel Zelaya took refuge, broadcast loud music in alternation with the alarm.[97]

The United States also has used it at borders and in prisons: "inside the cell block, [the noise] was horrendous," Carl Gruenler, the company's vice president for military operations, said approvingly.[98] GIs used it in Iraq, not only to bombard Fallujah with hard rock but in warning mode, as a "pre-lethal" weapon. Sergeant Major Herbert Friedman indicated, for instance, that "the LRAD has proven useful for clearing streets and rooftops during cordon and search, for disseminating command information, and for drawing out enemy snipers who are subsequently destroyed by our own snipers."[99] In other words, Iraqi snipers were trying to protect their ears and flee the zone, which facilitated their elimination by US soldiers.

Technically, the LRAD is described as a "device" not a weapon.[100] If its maker tempered its initial metaphors, it's because this allows distributors to sidestep the U.S. and European prohibitions on weapons sales to China that have been in place since the Tiananmen Square massacre in 1989.[101] It allows law enforcement to avoid systematically asking for authorization or approval before purchasing the LRAD, and to avoid seeking out independent expert advice on its effects.[102] Finally, it allows the LRAD Corp. to publish glowing notices after its products are used to distribute information to survivors of natural disasters, such as in Haiti, or to counter anti-capitalist protesters in Canada.

Before the G20 in Toronto in June 2010, the police sent releases to the media urging that they characterize the LRADs purchased for the occasion not as weapons but as "mass communication tools." The Canadian Civil Liberties Association, which was skeptical, called for the devices to be prohibited based on the pain they inflicted and the risk of deafness, as well as on the fact that the alarm would affect children or passersby indiscriminately, and that many people might not come to the protests for fear of these new weapons—a manner of dispersing a protest before it even takes place, in a sense. A court authorized only the megaphone function and prohibited the use of the warning mode if law enforcement would not adopt stricter regulations. The Toronto police, which managed to negotiate the purchase of four LRADs for €23,000[103] and had intended to keep them after the G20, acceded partially to the injunction (fixing limits on use, but without conducting all the necessary preliminary tests and consultations), thus allowing law enforcement to use the warning mode for a duration not exceeding five seconds if the target was more than 10 meters away, and to use the maximum volume if it was more than 75 meters away.[104]

Various programs also aim to use LRADs jointly with other weapons. Project Sheriff, launched by the Pentagon in 2004, aims to "develop and rapidly field a series of operational prototypes integrating directed energy and kinetic systems, both lethal and non-lethal . . . Ultimately, however, Sheriff is intended to advance a prototype of combined physical and psychological effects, as opposed to simply combined arms, on the tactical battlefield."[105] The idea, in particular, would be to combine the LRAD not only with the system of classic weapons but also with the Active Denial System (ADS) developed by Raytheon, which emits microwaves that produce a sensation of intense burning on the skin. Another weapon under development in the Applied Research Laboratory at Penn State, the Distributed Sound and Light Array (DSLA), would combine the LRAD with a blinding light.[106] On the acoustic side, the JNLWD indicates that the DSLA could alternatively use an "acoustic projector Thor-16S" (target high-output responder), capable of distributing low frequencies in a directional manner.[107] In short, according to the project director in 2004, Colonel Wade Hall, the Sheriff "is there to keep the peace, but has the option to go to destruction status if he needs it."[108]

### The Current Flourishing of Warning Devices

The LRAD Corp. is not the only maker of security loud-speakers, but the omnipresence of its device on navy ships has in fact elicited a complaint—which was quickly dismissed—about the equity of the tests conducted by the army. The plaintiffs were its competitors: IMLCORP, the maker of the SoundCommander, and Wattre Corporation, which proudly displays on its site the certificate accorded it by Guiness World Records in 2007, recognizing its Hyperspike HS-60 as the most powerful acoustic hailing device, with an amplitude of 140.2 dB at 128 meters. Wattre Corporation has in addition developed hypershields for the police, "portable sound cannons" that can broadcast voice or a "less-than-lethal" warning of 2,000 Hz at 140 dB at 1 meter.[109]

Among other warning devices are the Magnetic Audio Device (MAD), made by HPV Technologies,[110] and the Banshee II, tested by Lee Bzorgi, director of the National Security Technology Center of the nuclear weapons factory Y-12 in Oak Ridge, Tennessee. The Banshee II, which distributes sound at 144 dB, "has a frequency-switching system that pumps your ear drums, so it sounds like there's a drum beating there."[111] It presents the twofold advantage of being handheld and having a low cost.[112] The merchants of "sonic nausea," "sonic devastators," and the "next generation of alarm sirens" that cause people "to immediately modify their behavior" are legion, and their devices are used in both civil and military settings. For example, the Inferno Intenso, developed by the Swedish company Indusec, is used by Russian nuclear weapons depots, but also, depending on the model, by stores, schools, taxis, and offices.[113] The democratization of sound weapons is on the rise.

In the undersea world, the acoustic deterrent device, which since the 1970s has used high frequencies against marine mammals to protect fish farming, inspired the development since 2005 of technologies targeting humans.[114] The "integrated anti-swimmer system" used by the U.S. Coast Guard combines acoustic sensors that permit the detection of a presence with a powerful loudspeaker. Another prototype of a "diver interdiction system" is, moreover, financed by the JNLWD.[115]

China, which acquired the LRAD in 2007, has since developed its own device, which claims an amplitude of 146 dB at 1 meter.[116] During the Idex armaments fair, which took place in Abu Dhabi in February 2011, the Chinese company CETC International vaunted the merits of its "Directed High-Intensity Acoustic Low-Lethal Weapons for Police" with an exuberance reminiscent of the first developers of sound weapons: "By stimulating the human hearing sense, internal organs and central nervous system and other organs, it can weaken or destroy the mobs' hearing effectiveness so as to control the situation."[117]

In 2004, the Israeli army, for its part, put into action a weapon that was carefully shrouded in mystery. Even its name varies: the Shout, the Shriek, the Scream. According to the rare information put out by the military, it induces "nausea and dizziness."[118] Journalistic accounts speak of a machine similar to the hypothetical British squawk box, combining two low frequencies to affect the inner ear: "the knees buckle, the brain aches, the stomach turns. And suddenly, nobody feels like protesting anymore."[119] On a more prosaic level, the "non-lethal high-power acoustic radiator" made by the Israeli company Electro-Optics Research & Development Ltd. (EORD) is nicknamed "the Shofar," referring to the ritual trumpet, and combines, on a principle apparently close to that of the LRAD, thirty-six acoustic boxes functioning in phase to reach a claimed amplitude of 128 dB at 50 meters.[120] The Palestinians and their supporters, accustomed to confronting more aggressive weapons, speak neither of low frequencies nor of devastating effects but of a "powerful siren sound . . . just very noisy," and prepare for it by plugging their ears with cotton.[121] One soldier expressed enthusiasm: "It is probably the cleanest device we have ever had"; a human rights activist commented sarcastically that the Israeli power no doubt wanted to "disinfect the conflict."[122]

According to the journalist Hacène Belmessous, the French army considered equipping itself with Shofars and tried unsuccessfully to have them tested by the gendarmes at a training center in Saint-Astier. The gendarmes were less than eager to act as guinea pigs for a product whose harmlessness was in question. The director general for weaponry recognized in effect in 2008 that "we would expect numerous wounds," that "it would be unbearable for the residents of towns and cities," and "the capacity of our citizens to accept such a tool seems difficult to envision"—but recommended nonetheless "following the evolution of the matter."[123] In 2009, a journalist for *Le Figaro*, writing about the use of the LRAD during the G20 protests in Pittsburgh, noted that in France "a working group bringing together the police and the army was recently created to evaluate the benefits and dangers of acoustic weapons."[124]

### The Mosquito

Other sound repellant devices are already present in France, such as the Mosquito, which was invented by Howard Stapleton and marketed by the British company Compound Security Systems (CSS) in 2005. Drawing inspiration from ultrasonic aversion devices used to keep mosquitos, dogs, rats, or birds away, the Mosquito emits a high-frequency sound and thereby plays in daily life the same role that the LRAD plays in conflict situations, which is to say that it produces an unbearable sound to distance a category of people from a given location. Mosquitos are not weapons, strictly speaking, and they are not used by the police or military: they are, on the contrary, part of a private and commercial initiative that aims to establish preventive, permanent, and vigilant control of the public space, allegedly to supplement gaps in law enforcement—but in reality to intervene well beyond the realm of legal action. Indeed, according to the manufacturer, the "police and ASBO's (anti-social behavior orders) have only limited success in dealing with these problems."

The system, which is sold in the form of a cube around 12 centimeters per side and attached to a wall, reaches an amplitude of 95 to 108 dB with a range of 40 meters, and can function in two modes: "adolescent" or "all ages." At frequencies between 17,000 Hz and 18,600 Hz (very high but not ultrasonic), it generally disturbs only those who are under the age of twenty-five, since one of the characteristics of human hearing is that as we age, we no longer hear very high frequencies. CSS lists control of the homeless as one of the Mosquito's applications, indicating that "many police forces around the world have asked us to produce a Mosquito that affects older people too." Therefore, the device also offers an 8,000 Hz frequency, which is audible and irritating for the entire population. While below the threshold of pain, it is sufficiently unpleasant that between "five and fifteen minutes" of exposure is sufficient to force someone to leave the zone, thus presenting a solution to "vandalism," "graffiti," and "loitering" as well.

For the company, "the conventional definition of anti-social behavior notwithstanding, sometimes it is just not acceptable for a group of hooded teenagers to be hanging around blocking entrances and exits to shops, banks, etc., as it puts off customers from approaching the building." Thus the law that allows anyone to circulate freely in the public space, to be at leisure and to talk, may be an obsolete "convention": the Mosquito, sold on the Internet for €680, allows one to authorize only "social behaviors" compatible with a comfortable capitalism. CSS, which has thought of everything, also sells a Mosquito alternative: a "music player/mood calming system" that plays "royalty free classical or chill-out music to deter or discourage anti-social behavior."[125] This technique was used in the 1990s in Australia by shopping malls and municipalities, which broadcast songs by Bing Crosby, Brahms, Mantovani, and other "old fashioned music" to send adolescents fleeing—"pink lights" bringing out adolescent acne was the final blow.[126] It is, more generally, the principle of Muzak.

The Mosquito, which was welcomed by police stations, businesses, mass transportation authorities, British schools, and American high schools, has nonetheless aroused strong public opposition:[127] in 2008, a campaign entitled Buzz Off demanded its interdiction in Great Britain. Launched by the pediatrician and children's commissioner Albert Aynsley-Green, it criticizes both the demonization of youth inherent in the device, and the exposure to these high frequencies of young children and babies, whom their parents cannot protect since they themselves don't hear the Mosquito.[128] The legality of the device is further questioned in Ireland, since it does not respect the Non Fatal Offences Against the Person Act of 1997, which makes it a crime to expose others to any force, whether the application of "heat, light, electric current, noise or any other form of energy," without legal justification and without consent.[129]

In France, the Mosquito was offered for sale in 2006 by a company called IPB under the name Beethoven. "A sound that softens behavior," said the website slogan, before the site disappeared. Although it immediately seduced property managers, neither the public, politicians, nor the legal system proved favorably disposed. Despite the strategy of discretion adopted by IPB and its clients, in 2008 the court of Saint-Brieuc prohibited its use by a private citizen in Pléneuf Val-André, arguing that it would cause an "abnormal disturbance to the neighbors" and be bothersome to everyone's hearing. The court based itself in particular on the public health code, which bans all noise that "by its duration, its repetition or its intensity is liable to disturb the tranquility of the neighborhood or human health, in places public or private."[130] The authorities were alerted by the Val Tonic business association, which called it an "illegal sound weapon" and, more originally, a "sonar repellant." Neighbors complained of migraines, nausea, and tinnitus, and children plugged their ears when passing the house in question.[131]

In a similar vein, the device provoked a lively debate in Belgium, where it was banned by certain municipalities. In Great Britain, doctors and citizens worried about the health effects, but British studies on the matter were inconclusive, and the Home Office qualified its effects as essentially "subjective." Europe did not come out in favor of banning Mosquitos either, and only rapid and local mobilization managed to silence some of them. The box therefore continued to make its way commercially, proclaiming loud and strong its harmlessness and perfect legality.[132] A survey of the health limits set by various countries regarding exposure to high frequencies and ultrasounds transmitted by air nonetheless indicates values well below the potential of the Mosquito: for frequencies of 16,000 Hz or 20,000 Hz, a maximum threshold of 75 dB is indicated for eight hours of exposure, and 85 dB for less than an hour.[133]

## "GHOSTS SOUNDS": THE SEARCH FOR THE SUBLIMINAL[134]

In recent years, another device has popped up alongside the Mosquito: the ultrasonic loudspeaker, which can precisely target a person or a zone of emission. The ultrasounds used abundantly in the medical and industrial fields find one of their rare known military applications here: they enable an audible sound to be carried a long distance solely toward designated auditors, whether to incite or to repel. An audible message is first converted into ultrasounds, which are inaudible, long-range, and very directional, and which reverberate off obstacles. When the ultrasonic wave is distributed by loudspeaker, the slowing and distortion it experiences passing through the air restore the initial frequencies of the message, and thus its audibility. The sound can thus be "directed" (targeting a zone) or "projected"—the wave in this case is sent over a surface that reflects it and changes its angle of diffusion, making identification of the source more difficult.

In the early 2000s, two comparable systems using this technology went on the market, their respective inventors facing off to gain the favor of the media: on one side was Elwood Norris of ATC Corporation (the future LRAD Corp.) with its Hypersonic Sound System; on the other was Joseph Pompei, of the company Holosonics, with its Audio Spotlight. Like combat loudspeakers and the LRAD, these ultrasonic loudspeakers are hybrids: half civil technologies, half weapons, at times tools of communication, at times used for coercion. Their invention is in fact a revolution in the acoustic world, since it allows two people in the same room to listen to different music, for cinemas to distribute sound in a more realistic fashion, for museums to put away their CD- or cassette-based audio guides. The Bibliothèque nationale de France, notably, acquired Audio Spotlights beginning in 2001; some artists and musicians are interested in the possibilities, and according to Elwood Norris, "Disney is nuts about it!"[135]

The ultrasonic loudspeaker opens such promising possibilities that in 2004 Holosonics was among twenty-five companies invited to present their new wares at the great capitalist fiesta of the G8 in Georgia. "Add sound and preserve the quiet" was Holosonics' slogan: a perfect approach. The main users of the ultrasonic loudspeakers are, for the moment, stores, which find in it a new, more effective, and endlessly reproducible manner of attracting customers: the message can be broadcast on the sidewalk in front of the store, it can appear to be coming from a product on the shelves, or it can be emitted from a distance of 200 meters.[136] One of the first advertisements to use it was for a televised series dealing with paranormal phenomena; broadcast from the roof of a building in New York in 2007, it whispered in the ear of passersby, "Who's there? Who's there? It's not your imagination."[137] The

ultrasonic loudspeaker accomplishes an old dream of marketing creative departments: insinuate oneself directly into the
thoughts of the consumer. That at least is the fantasy the first
advertisers who used it expressed, for once it was a known and
widely used technology, it would no longer surprise anyone—
but would preserve its capacity to exploit the slightest gaps in
the public sound arena.

By enabling one to choose who will hear a sound and who
will not, the ultrasonic loudspeakers can induce different behaviors in the same space: that is their interest as a technology of control and surveillance. They can in effect be used in
factories to "broadcast information while allowing the worker
to remain concentrated on his work," or to serve as "repel-
lents against pigeons . . . or . . . against homeless people,"
according to Directional Sound, one of the European import-
ers of Audio Spotlight.[138] U.S. law enforcement agencies are
now among the clients of Holosonics and of LRAD Corp.:
the loudspeakers serve both as a means of discrete communication and as tools of deception and disorientation. The Los
Angeles police department "wants to try it on high-crime al-
leys,"[139] the navy equips some ships with them, and the army
mentions the possibility of sending "ghost sounds [that] could
be bounced off 'rocks or any reflective surface' to fool people
into believing they were not alone."[140]

The Defense Advanced Research Projects Agency
(DARPA) has also been working since 2007 on the production
of its own "sonic projector," which would serve as "a method
of surreptitious audio communication at distances over one
kilometer." It notes that "the sonic projector system could be
used to conceal communications for special operations forces
and hostage rescue missions, and to disrupt enemy activities."[141] Other uses can be envisaged, such as sending an order

to a person in the middle of a crowd without anyone nearby hearing a thing.

We should note, finally, that certain non-acoustic electro-magnetic devices attempt to achieve a similar effect (though one that is less precise, less reliable, and prone to disturbing secondary effects) by exploiting an effect called "microwave hearing." A document put out by the French Institute for Research and Security (Institut national de recherche et de sécurité), seeking to establish exposure limits, describes the phenomenon: "People whose hearing is normal can perceive fields of pulses-modules at a frequency between 200 MHz and 6.5 GHZ [radio frequencies but not acoustic frequencies]. . . . The auditory effects of microwaves has been attributed to a thermoelastic interaction at the level of the auditory zone of the cerebral cortex."[142] According to the report "Bioeffects of Selected Non-Lethal Weapons," declassified in 2006 by the U.S. Army, "Microwave hearing is a phenomenon, described by human observers as the sensations of buzzing, ticking, hissing, or knocking sounds that originate within or immedi-ately behind the head. There is no sound propagating through the air like normal sound. This technology in its crudest form could be used to distract individuals; if refined, it could also be used to communicate with hostages or hostage takers di-rectly by Morse code or other message systems, possibly even voice communication."[143]

Research on microwave hearing, which began during World War II, led in particular to the development by the U.S. Navy of a weapon called MEDUSA (an acronym for "mob excess deterrent using silent audio") in the early 2000s, a weapon to dissuade riots by means of silent sound. But no reports of progress were published, and biophysicists were skeptical, since "the high power outputs required to transmit sufficiently loud sound levels would heat the brain, causing tissue damage and death fairly rapidly."[144]

Unlike lower frequencies, music as well as medium and high frequencies required no technological advances to be exploited by the police and the military. But over the course of the twentieth century, technology introduced a new form of domination, both more distant and more intrusive: sound is used to express power and to domesticate rebel bodies and minds. Technology also allowed for expanded, omnipresent control over time and space: sirens, sound aversion devices, and directional loudspeakers create a hygienic map, breaking space into zones, some authorized, some forbidden, with prescribed behaviors in each location. Boundaries have become blurry between the civilian and the military, between the entertainment industry and the military-industrial complex, between consumption and repression, between the private and the public spheres.

Industrial and military development today is mainly focused on middle and high frequencies because they create the possibility of managing individuals in the social space via sound—at first intuitively, then in an increasingly rationalized manner. As in the case of sensory deprivation, nothing is flagrant, and everything is subject to infinite legal quibbling; we slip almost imperceptibly from an accepted, inoffensive, or entertaining use of medium and high frequencies to a potentially dangerous, deliberately aggressive, even "pre-lethal" usage. Sound is a gray zone that seems to allow for any and all experimentation.

# 6

## "NO MATTER WHAT YOUR PURPOSE IS, YOU MUST LEAVE": THE SOUND OF POWER[1]

What does the development, in recent times, of the use of sound as a means of control and repression mean? Having reviewed the inventory, we know that the research is already more than half a century old, that it was laborious, and that it produced only a few devices and specific uses, despite many long years and high hopes. That said, today we are nonetheless witnessing the increasing presence of repellent sound technologies being used in the public arena. The phenomenon is recent and the debate it occasions remains limited; therefore, we can only sketch out a preliminary analysis here.

We shall see, first of all, that sound used by the military, the police, and some private citizens sheds remarkable light on the genesis, proliferation, and art of camouflaging "non-lethal" technologies. But all this may already be ancient history: beyond the "non-lethality" of the technology, these weapons are becoming *entertainment*, to borrow the analysis of philosopher Olivier Razac. We will look here at the growing interaction between the entertainment and weapons industries: sound is used here to scramble the boundaries between war, culture, and games, and to cleanse public debate of all political content. As for the future, it seems to go beyond military-entertainment, or rather to imbed it in the very places we inhabit: we shall see that the acoustic space, a terra incognita, is progressively being conquered, becoming the realm of an omnipresent social triage, progressively determining the contours of a new sound urbanism.

## "GET A HIGH-POWERED MEGAPHONE, NOT AN AUDITORY MACHINE GUN": "NON-LETHALITY"[2]

A member of the U.S. Congress, James Scheuer, commented enthusiastically in 1970, "We can tranquilize, impede, immobilize, harass, shock, upset, stupefy, nauseate, chill, temporarily blind, deafen or just scare the wits out of anyone the police have a proper need to control."[3] This recalls, in less cheerful terms, the definition given by philosopher Michel Foucault of biopolitics: "this great double-edged technology—anatomical and biological, individualizing and specific, turned toward the performance of bodies and looking toward the processes of life—characterizes a power whose highest function now may no longer be to kill but to infiltrate every aspect of life."[4] It is no longer a matter of eliminating the enemy body, but of rendering it docile or ineffective, provoking its total paralysis.

### Genesis

If "non-lethality" began to be formulated as such only in the 1960s, "in a context marked by the emergence of mass protests and civil rights movements," and if it became a central strategic concept only in the course of the 1990s, the history of the use of sound as a weapon brings to light older strata of the same idea.[5] Before Colonel Rex Applegate advocated the use of blank grenades for law enforcement, before researchers and the military worked to find the miraculous infrasound that would produce "incontinence" and "lethargic states," before "dissonant sounds" were used to disarm Vietcong marksmen or Western demonstrators, sound was already being used in "non-lethal" experiments. Current weapons and the ideology that underlies them are at the crossroads of several intertwined histories.

THE FRONT LINE: the psychological operations of the army. This is where a certain conception of "non-lethality" was formed, envisaged as a complement to, accompaniment to, and sometimes intensification of lethal action. Following the precept of Sun Tzu, a sixth-century B.C. Chinese general abundantly cited in the psyops literature, it is a matter of "subjugating the enemy's army without doing battle";[6] as a U.S. officer wrote in a letter in 1947 with regard to the term "psychological operations," there is a "great need for a synonym which would be used in peacetime that would not shock the sensibilities of a citizen of democracy."[7] These are precisely the three characteristics of "non-lethality": the goal of neutralizing rather than killing, the creation of blurriness between war and law enforcement or humanitarian operations, and the taking account of the media and public opinion in the management of conflict.

SECOND LINE: sensory deprivation. Starting with World War II, research in this domain and the boom in psychological operations would slowly institute a science of behavior and of sound conditioning, unintentionally at first, then with growing coordination. Doctors and psychiatrists were introduced into military operations, not to provide care but to make these operations more effective and less subject to reproach. The collaboration, begun in secret with Project MKUltra, became official for the "enhanced interrogation" branch with the presence of behavioral science consultation teams in CIA prisons and for the "psychological action" branches with the behavioralist laboratories of ARDEC, such as the Target

Behavioral Response Laboratory (TBRL), which is engaged in "national defense,"[8] and the Stress and Motivated Behavior Institute (SMBI), which specializes in neurobiology and brings together "a multidisciplinary team of researchers and theorists to collegially tackle problems related to stress and coping with an eye toward our nation's interests."[9]

*THIRD LINE:* management of living beings. A fair number of "non-lethal" devices were initially developed to control insects and animals. This is the case of "infrasonic barriers" to keep away fish, "scare cannons" to frighten birds, "repellent acoustic devices" to ward off marine mammals, and "ultrasound generators" to ward off insects, rodents, or dogs—not to mention, beyond the realm of sound, electric fences or prods to control cattle, which led to the invention of the Taser.[10] A special rhetoric was established to reverse the logic: the version of the LRAD that aims to keep birds away from airplanes or fields, for instance, is presented as "protecting wildlife."[11]

In more prosaic terms, "non-lethal" devices do not so much protect their targets as their users: no doubt we save a few birds from the turbines, but the goal is to protect the industry and transportation rather than nature. When "non-lethal" practices begin to be used against human beings and to orient their activities, they too aim not so much to spare humans but rather to domesticate or break them. Initially, they are tested on humans acting as guinea pigs, who are sometimes not volunteers. Then they are directed against populations the authorities consider not the most vulnerable but, on the contrary, the most harmful, the most dangerous, or the least "civilized": armed revolutionary militants and dissidents on the home turf, colonial subjects, anti-imperialist combatants, and terrorism suspects abroad. The British researcher and pacifist Steve Wright has noted that "non-lethal" weapons "have featured regularly in human rights abuses since they first found a role in crowd control operations in colonial times."[12]

*FOURTH LINE:* The institutional forecast. In "L'Arme non létale dans la stratégie militaire des États-Unis" (Non-lethal weapons in U.S. military strategy), researcher Georges-Henri Bricet des Vallons reviews the history of "non-lethality" from the perspectives of legality and resources, and from law enforcement and military points of view. The National Institute of Justice and ARDEC were interested in the "development of incapacitating weapons to control civilian, mainly incarcerated, populations," while the Department of Energy was interested in devices capable of protecting nuclear sites.[13] Meanwhile, the American Correctional Association and the National Association of Sheriffs see them as a means of improving law enforcement. In 1994, "following the riots in Los Angeles in 1992 and the tragedy of the siege of Waco," the Departments of Justice and Defense signed an agreement to share technologies that can assist "operations other than war" and law enforcement operations; DARPA is charged with global supervision.[14] Law enforcement becomes militarized and military action takes on the aura of law enforcement. In the late 1990s, the NIJ signed cooperation agreements on "non-lethal" weapons with Great Britain, Israel, and Canada; NATO and Europe then began heading down the path paved by the United States. The use of "non-lethal" weapons, initially conceived as limited, has expanded to become the norm in law enforcement and military operations.[15]

### Proliferation

Thus "non-lethality" became "central to military thinking on asymmetric conflict and urban warfare." Asymmetric conflicts—conflicts between opponents whose organizations, objectives, means, and strategies differ significantly—face different criteria, given "the permanent presence of media in the conflict, the growth in the number of prolonged urban conflicts . . . the dilution of the semiotic distinction between the military and the civilian, between combatants and non-combatants, between military and paramilitary (private military companies)." War has become a "cross-cutting system of control and a continuum of force"; the army has responded to this transformation by officially declaring a "revolution in military affairs," which aims notably to "develop a panoply of new weapons . . . based on lasers, acoustic waves, electro-magnetism, super-adhesive materials, super-caustic materials, etc." A world of possibilities opens that is "so broad that it touches on all techno-scientific branches imaginable and thus offers a particularly lucrative market to the military-industrial complex."[16] Initially conceived thanks to the encouragement and financing of the military, "non-lethal" products would come to be offered and sold on the open market by their manufacturers.

Two forces are at work simultaneously: the realm of war is both spreading and becoming less visible. This is evidenced notably by the appearance in daily life of weapons initially conceived for the battlefield: the "civil-military duality . . . lies at the heart of the concept of non-lethality."[17] The evolution of loudspeakers in conflict zones is a case in point. First used on the battlefield as tools of communication, they were later used in deception operations, then in harassment aimed at dislodging or killing the enemy. As the concept of the "enemy in our midst" evolved, these same loudspeakers were integrated progressively into police missions, as seen in the case of the LRAD. In a second-phase extension of the domain of war, weapons used by law enforcement were duplicated in the civilian sphere by devices of narrower range, conceived outside any program of state support—the Mosquito, used by businesses, local governments, or individuals, is the best example. The most recent inventions, such as ultrasonic loudspeakers, are directly conceived for dual use, military and civilian, and offer the functions of both communication and coercion—they are devices that simultaneously suit demands for advertisement, entertainment, communication, and repression.

Before the expression "'non-lethal' weapons" gained currency in the media, these weapons were called "rheostatic," a rheostat being a device that allows one to modulate the intensity of an electric current passing through a circuit. The term, used in official U.S. terminology during the 1990s, characterized more precisely the goal of these weapons: to offer an adjustable effect ranging from "non-lethality" to lethality. In the area of sound, the various efforts by the company SARA furnish evidence of the research that has been done. The acoustic technologies currently deployed, while not a priori deadly—the result more of a failure in the research than of a voluntary restriction—nonetheless respond to this definition since they allow one to pass from "safe" usage (symbolized by a green area on the LRAD's volume control) to a dangerous use (shown by a red zone). The latter level is dangerous for hearing, but dangerous as well because it deprives an individual of the means to defend himself and thereby makes it possible to eliminate him, as was the case of the militants ambushed in Iraq: the "non-lethal" weapon served in this case to augment the lethal capacity of classical weapons.

These devices are also rheostatic in their field of application, which extends from the battlefield to personal self-defense. Their hybrid status, between weapon and civilian technology, favors their sale, which is on the open market; as we have seen with the LRAD, no official permit is needed to use them. This allows activists to fight with equal weapons, at least on the acoustic level, against the authorities or against private security teams. Thus the ecologists at Sea Shepherd equipped themselves with an LRAD to respond to the whalers they were chasing. But the "innocuous" nature of these technologies also allows the armed forces to be free of the "proportionality" and "discrimination" requirements that constrain the use of weapons.[18] Various independent observers speak of a "danger of proliferation";[19] they note that criminal use of these weapons is increasing and fear that the technologies may also serve as new instruments of torture, notably in authoritarian countries—the risk being all the greater as the manufacturers minimize their potential for injury.[20] "Non-lethal" technologies therefore raise not only the threshold of authorized violence legally exercised by the armed branches of authority but also the threshold of violence in the society at large.

### *Camouflage*

As it becomes more diffuse, war becomes less visible—it is internalized. The battlefield progressively moves to inside the individual and the external traces become more discreet. In a 1998 article entitled "The Mind Has No Firewall," U.S. Army lieutenant colonel Timothy L. Thomas advises us to consider the human organism "just as any other data processing system" in the "information war," whose "psychological and data-processing capabilities" can be altered in order to "confuse or destroy the signals that normally keep the body in equilibrium."[21] This implies a hygenicized conception of war,

war that is "clean"—immaterial and total. The body of the enemy, which previously carried traces of aggression and thus bore witness, is now conceived as an envelope to hide wounds, like the facade left intact in front of an interior that is demolished or destroyed. The "non-lethal" weapon causes a scandal only when it leaves mutilation visible.

In 1975, military researchers noted that "it is preferred that onlookers not get the impression that the police are using excessive force or that the weapon has an especially injurious effect on the target individuals. Here again, a flow of blood and similar dramatic effects are to be avoided."[22] The energy used is invisible, as are its outcomes: no need to clean the streets of the city after an acoustic battle, nor to worry about protests on the part of its victims. One may manage to prove that a wound was caused by a club or a Flash-Ball, but it would be far more difficult to blame an LRAD or a Mosquito. It's easy to point to a visible nuisance, but far more challenging to point to a phenomenon that only a part of the population can even perceive.[23] The difficulty is compounded by the fact that independent medical research on acoustic weapons remains rare and publications are sparse, while the psychological effects are little accounted for, except from a strategic point of view. As Steve Wright sums it up with regard to these weapons, "A key design criterion is that they should appear rather than actually be safe."[24]

But while generally neither visible nor lethal, the wounds inflicted by these weapons are nonetheless real and the obligation to respect international conventions on weapons no less relevant. Four warnings may be recalled. Jürgen Altmann provides a technical recommendation, suggesting stricter limits on the amplitude authorized for law enforcement: a limit of 120 dB would take account of individual sensibilities as well as of the diversity of situations on the ground.[25] We might consider that 120 dB is still too high and study the medical recommendations made by British doctor Robin Coupland, editor of the Red Cross' Project Sirus. The study, published in 1997, aimed to determine "which weapons cause 'superfluous injury' or 'unnecessary suffering.'"[26] Such weapons are essentially banned by international law, but the definition remains too vague to be efficacious. More precisely, Project Sirus proposed to banish any weapon that would cause a "specific disease, specific abnormal physiological state, specific and permanent disability or specific disfigurement"[27]— criteria that, were they applied, would significantly restrain the current proliferation of acoustic, biochemical, kinetic, and directed-energy weapons.

The problem remains as to which authority has the power to enforce—or, in this case, not enforce—the criteria. Sociologist Brian Rappert, in an article entitled "A Framework for the Assessment of Non-Lethal Weapons," remarks in particular that "police and military forces should not be left alone to police themselves regarding the evaluation and control of weaponry," that "manufacturers' statements of the efficacy, purpose and safety of a weapon should not be taken on faith," and, more broadly, that "alternatives to new force options should be pursued first."[28] If we're not happy with a simple framework for containing authorized violence, we should consider a political warning, one formulated by Steve Wright, who laments the fact that "non-lethal" weapons claim to offer "an apparent technical 'fix' for wider social and political problems."[29] The question is no longer so much the regulation of these weapons as their existence and their proliferation in the first place outside of any public debate—even, one might say, instead and in place of any public debate.

## "TALES OF SYMPHONIA": WEAPONS FOR ENTERTAINMENT[30]

The public debate hasn't exactly gone away: it is swollen with a carefully cultivated absence of content, transformed from the inside to occupy the space in the most inoffensive manner possible. The second Gulf War—or rather its televised representation—would play a decisive role in the emergence of "non-lethality." The broadcasting of images of dead Iraqi soldiers by the American networks would shock public opinion to the point that Marine lieutenant colonel Duane Schattle would speak of the necessity of taking the "CNN factor" into account from then on, which is to say the mediatization of war.[31] "Non-lethal," media-friendly weapons were the answer to this new reality. They go over well on television and thus increase military options: "The advantage of such a weapons system is obvious: from a politico-military point of view, they provide an argument to legitimize operations that could not have been legitimized with conventional weapons, while from a tactical point of view, they offer decision-makers on the ground a supplemental option for intervention."[32] What's more, representation and entertainment are inextricably combined today with "non-lethality"—they are intertwined, they transform it. They make it seem almost like an old, accepted idea, the presupposition on which the war of the future—and daily life—will be based.

### Torture on the Playlist

The resurgence of sensory deprivation during the "war on terror" is a flagrant example of this evolution. Suzanne Cusick takes issue with the euphemism "no-touch torture" that is habitually used to describe it, due to the way in which it masks the violence. She notes the same type of minimization in the manner in which this torture was discussed in the blogosphere when the first information on Guantánamo was made public. "Is it torture?" and "What's the playlist?" were to her mind the two questions around which most of the online commentaries revolved. And many were the jokes: proposals for different playlists, talk of fights with the neighbors, and their musical tastes.[33] In other words, music offers the advantage, to those who use it as a weapon, of clouding the debate: torture becomes funny, socially acceptable, and telegenic.

James Hetfield, from the group Metallica, stated in 2004: "We've been punishing our parents, our wives, our loved ones with this music forever. Why should the Iraqis be any different? . . . For me, the lyrics are a form of expression, a freedom to express my insanity. If the Iraqis aren't used to freedom, then I'm glad to be part of their exposure."[34] Four years later, Bob Singleton, musical director of *Barney*, dismissed the idea that the song "I Love You" could be used as a weapon: "It's absolutely ludicrous. A song that was designed to make little children feel safe and loved was somehow going to threaten the mental state of adults and drive them to the emotional breaking point? . . . The idea that repeating a song will drive someone over the brink of emotional stability, or cause them to act counter to their own nature, makes music into something like voodoo, which it is not."[35] As we have seen, there is no magic and plenty of science in the fact of breaking someone via sensory deprivation, whatever the specific content. The magic seems instead to lie in the fact of making torture disappear, dissembling it as an entertainment product.

This form of torture is for Cusick perfectly representative of totalizing power as it is exercised today. When music is used as a weapon, it gives the sense of "touching without touching": "I imagine it, sometime, as being plunged into it something like the post-modern, post-Foucauldian dystopia where one is unable quite to name, much less resist, the overwhelmingly diffuse Power that *is* outside one, but also *is* inside, and that operates by forcing one to comply against one's will, against one's interests, because there is no way— not even a retreat to interiority—to escape the pain."[36] Music becomes the instrument of a total domination of the body and the spirit, it is charged with proclaiming the all-powerfulness and omniscience of the authorities, as well as the futility of all resistance. Psychologist Françoise Sironi sums it up this way: "In fact, we torture in order to silence. Torture reduces the torturer and his victims to the same silence."[37] And it silences the spectators as well—or makes them chatty about anything that will distract.

### Imagination as an Instrument of Domination

Not only does sound allow one to create bubbles of silence and forgetting, but its military or police usage makes manifest an instrumentalization of the imagination. The difficulty in understanding the functioning and effects of acoustic weapons, as well as the mass of conspiracy theories and paranormal inventions they inspire, works in their favor: the information about them becomes confused, thus fueling the psychological effect from which they benefit. The invisibility of the "incapacitating" energy they emit and the difficulty at times of becoming aware of their use makes them difficult to grasp, immaterial and mysterious. While infrasounds are for the moment nonexistent in law enforcement and while ultrasounds are little used, both participate fully in the world of the imaginary—even fueling it more than devices commonly in use. Weapons of high technology that, like "no-touch torture," touch without touching, pass through obstacles, and act without seeming to act, acoustic weapons are also infused with a carefully sustained illusion of magic, which allows one to keep the other at a distance and to manifest power. They fascinate and they subjugate.

Among the first promoters of "non-lethality" was an unusual team: "In the 1990s, science fiction writers and Quakers Janet and Chris Morris joined forces with futurologists Alvin and Heidi Toffler to advocate a new form of bloodless warfare."[38] Far from being dismissed by the military top brass, they became associated, along with Colonel John Alexander—"hard-core mercenary turned thanatologist"[39]— with the research conducted by a conservative think tank, the Global Strategy Council, then under the leadership of former CIA analyst Ray Cline.[40] In 1980, in an article entitled "The New Mental Battlefield," Colonel Alexander revealed his fascination—as in the early days of Project MKUltra—with Soviet advances in "paranormal applications," and sketched out a military-literary forecast:

> Certainly, with development, these weapons would be able to induce illness or death at little or no risk for the operator. Range may be a present problem, but it will probably be overdone if it has not been already. . . . No special psychic ability is necessary to charge the generator. The psychotronic weapon would be silent, difficult to detect and would require only a human operator as a power source.[41]

The presence of fantasy-mongers in the closed club of the inventors of "non-lethality" is not without significance: it reveals the imagination of the powers that be, marks the growing interest the army and the police have in communicating about their operations, and characterizes the emergence of what will progressively replace the military-industrial complex: the military-entertainment complex.

Rumors, as we see from reading the declarations of sound device promoters over the years, are first and foremost the business of manufacturers or militaries, eager to sell their products and become the pioneers of the weapons of the future. These same rumors are then taken up, detached from their official origins, by a part of the press or public opinion. Jürgen Altmann noted in 1999:

> Well-founded analyses of the working of NLW [non-lethal weapons], the transport/propagation to a target, and the effects they would produce, are urgently required. This holds all the more, as the published sources are remarkably silent on scientific-technical detail. Military authorities or contractors involved in NLW research and development do not provide technical information.[42]

There is no chance involved: during the first large military conferences on "non-lethal" weapons, U.S. advocates of their use concluded that "secrecy is of paramount importance to ensure maximum effectiveness."[43]

Since then the reality has hardly evolved concerning acoustic technologies. Initially, the enthusiastic inventors of the new weapons endowed their specimens with qualities as extraordinary as they were terrifying. In a second phase, some, notably in the United States, adjusted how they spoke about the devices, tempering their assertions and insisting more on the harmlessness of the devices and their suitability for asymmetric conflicts or for daily life. The newcomers to the industry of "non-lethality," such as China and Israel, continued to prefer picturesque metaphors to euphemisms. In either case, the weapon is protected by a system of military-commercial propaganda, by an image of itself that substitutes for the real thing: a way of making it visible while keeping it secret—and thus a way to "legitimize or mask their institutionalized use of force."[44]

### War as Recreation

"Sonic hatchet," "sonic sword," "sonic weapon and intense thunder," "sonic slicer," "sonic eruptor," "sonic storm," "sonic blow": sound makes for a marvelous weapon in video games—nothing like it to "devastate" and "destroy the enemy." With the Sonic Grenade and Sonic Grenade Pro apps for mobile phones, you can get "instant crowd control."[45] Less dramatically, the young have the means to call one another without adults being able to hear: shortly after the release of the Mosquito, a ringtone for portable phones was named after it, using the same frequencies but at a very low sound level for brief periods.[46] Browsing certain forums on "special grenades" and other "resonators," you can't tell at first if it is a matter of real weapons or of their virtual substitutes. While doubts are quickly dissipated, the confusion is symptomatic of the tight bonds between the weapons and entertainment industries. Media theoretician Friedrich Kittler strikes at the heart of the matter: "The entertainment industry is, in any conceivable sense of the word, an abuse of army equipment."[47]

A figure such as engineer Harold Burris-Meyer is particularly representative of the growing interaction between the two: a man who conceived, at various points, of the stereo system on which Disney based its Fantasound, the sonic deception of the "ghost army," the sound environment of the Muzak corporation, high-performance loudspeakers for military aviation, sound effects for cinema, a "sound torpedo" project—

and even a "psycho-galvanometer" aimed at measuring the emotional impact of sound on the public.[48]

The British company Decca, for its part, killed two birds with one stone: mandated by the Royal Air Force in 1940 to refine the capture of underwater sounds and make it possible to distinguish between British and German ships, it took the opportunity to improve its recording techniques. In the immediate post-war period, the company EMI (Electric and Musical Industries) took advantage of German military research into analyzing Morse code and recording: information furnished by the capture of German material for decoding allowed the company to create cassettes and tape recorders and to thereby outfit its Abbey Road studio for a quarter century.[49] Finally, Kittler reports that when "Karlheinz Stockhausen was mixing his first electronic composition, *Kontakte*, in the Cologne studio of the Westdeutscher Rundfunk between February 1958 and fall 1959, the pulse generator, indicating amplifier, bandpass filter, as well as the sine and square wave oscillators were made up of discarded U.S. Army equipment."[50]

Similarly, military research on the vortex during World War II would fuel the gaming industry. As we saw, Thomas Shelton's invention of a vortex launcher that could carry noxious gases was never used on the battlefield. On the other hand, Shelton "invented a toy pistol that would produce a ring capable of knocking over a paper target at a range of a few meters. This was the golden age of science fiction and ray guns, and the toy, dubbed the 'Flash Gordon Air Ray Gun,' was a big hit. *Popular Mechanics* voted it 'Toy of the Year' in 1949."[51] The compressed-air pistols would later be mass-produced and sold, from the 1965 Wham-O Air Blaster to today's Airzooka.

Of late, the relationship has reversed itself: the gaming industry has taken the initiative and no longer merely recycles

military material. A researcher at Stanford University, Timothy Lenoir, published a study in 2003 entitled "Programming
Theatres of War: Gamemakers as Soldiers," in which he describes the long-standing ties between the army and the entertainment industry, and their recent evolution:

> Indeed, a cynic might argue that whereas the military-in-
> dustrial complex was more or less visible and identifiable
> during the Cold War, today it is invisibly everywhere, per-
> meating our daily lives. The military-industrial complex
> has become the military-entertainment complex. The en-
> tertainment industry is both a major source of innovative
> ideas and technology, and the training ground for what
> might be called post-human warfare.[52]

The transition started for the most part in the late 1980s
with combat simulation programs put in place by DARPA.
To avoid unnecessary expenses, the U.S. military research
agency worked off of "technologies developed outside the
DOD [Department of Defense] such as videogame technology from the entertainment industries."[53] Pilots and soldiers
were thus able to start training in flight and combat under
"real conditions" via simulation platforms. In exchange, the
military brought its expertise to the game makers in order to
improve the realism of war scenes. Personnel exchanges, soldiers into entertainment and vice versa, served to formalize
the bonds.[54]

In 2009, the U.S. Army Research Laboratory opened its
Environment for Auditory Research, or EAR, a high-tech auditory complex built with numerous industrial partners, who
aim to study "the ability of soldiers to detect, identify, and lo-
calize sounds in realistic operational sound fields."[55] Over the

course of 2011, the Missouri University of Science and Technology converted a former warehouse into a surround-sound audio battlefield: sixty-four loudspeakers and four subwoofers broadcast recordings of M-16 assault rifles, Kalashnikovs, explosions, tanks, and helicopters to prepare soldiers for the sound environment of real war. The army has not yet approved the use of this acoustic simulator but is looking at it closely.[56] We should note that Steven Grant, who conceived of this "immersive audio environment" (IAE), vouches for its safety:

> With our current implementation of battle sequences you can stay in the IAE without any hearing loss for about two hours per day without using hearing protection. Right now we are generally running at a maximum of about 100 decibels. Of course, real battles' sounds are much louder, but we conform to OSHA (Occupational Safety and Health Administration) guidelines.[57]

A thesis published in 2005 by a doctoral student at the Naval Postgraduate School in Monterey, California, is entitled "Modeling Sound as a Non-Lethal Weapon in the COMBAT XXI Simulation Model." Its author, Joseph Grimes, explains that

> modeling non-lethal weapons was identified by TRAC-Monterey as important to better represent actual combat. This thesis used COMBAT XXI, a high-resolution, closed-form, stochastic, analytical combat simulation, to replicate non-lethals and study the effects on individual combatants. Existing source code was modified to model the Long Range Acoustic Device (LRAD), the non-lethal

platform chosen for this research. . . . Once the LRAD capability was developed, a scenario was developed to test the simulated effects of the device. . . . It was concluded that the implementation of this new non-lethal capability in COMBAT XXI improved the model and created a more realistic representation of actual combat conditions.[58]

Thanks to the LRAD, the virtual battlefield has gained in realism, while real war is more scripted.

This interconnection between war and games has had an effect on the acceptability and the use of "non-lethal" devices. The intermingling of wartime and entertainment applications serves to make such devices socially acceptable: ultrasonic loudspeakers, for example, can be used in an intrusive manner in the streets of cities or as "pre-lethal" weapons on the battlefield, but they will be highly appreciated in daily life for their qualities of isolation or acoustic depth. At the same time, the appearance of games that include "non-lethal" weapons is likely to influence the handling of such weapons by the police or military. In 2008, Olivier Razac published the results of his study as "The Use of Weapons of Temporary Neutralization in Prison." He mentions encirclement grenades in particular and warns against the "risk of sliding toward an abusive, even entertainment, usage of neutralization weapons due to the intrinsic ambiguity of this type of material."[59] He comes back to this question of ambiguity in an interview: "On the one hand, they must remain weapons—they must continue to be dissuasive, to scare people. But on the other hand, they are becoming banal. . . . That is the secret, I think, of these neutralization weapons: they make people scared and they make them laugh. . . . In this, they are disarming for critics."[60]

## A SONIC ARCHITECTURE OF CONTROL: TOWARD AN URBANISM OF SOUND[61]

At the end of his study on "non-lethal" weapons, Neil Davison concludes:

> Efforts to develop "non-lethal" acoustic weapons that have incapacitating effects have not been successful. . . . Given past assessments that rule out extra-aural effects below the threshold for permanent hearing damage, applications of these "hailing devices" appear limited to irritating or psychological effects and ongoing efforts to develop acoustic "non-lethal" weapons seem likely to fail.[62]

That the LRAD is physically less dangerous than the electric pulse gun, the Flash-Ball, and lasers is true. But we shouldn't underestimate the "irritating or psychological effects," which are non-negligible, as we have seen. In particular, acoustic devices, far from being insignificant in the evolution of law enforcement, brought about a new relationship to the public space, to hearing and those around us. They designate the first outlines of a hygienicized cartography of cities and border areas. They establish a new form of violence, more diffuse and more global.

In "Alarms and Sirens: Sonotypes of Daily Commotion," the researcher Noel García López, from the Autonomous University of Barcelona, states that "if sight is characterized by its distance from the object and its use of reason, hearing is collective and always passes for being the sense of emotions and affect."[63] Today, just like ten centuries ago, the sound environment is shared. The nature of sounds has changed, but our manner of receiving them is identical. What these acoustic devices of control are creating is a fragmented and functional

conception of the public space—everyone in his place, everyone in her zone. That implies totalizing technologies, which
administer orders that are impossible to contest, as all you can
do in the moment is obey the prohibition of access to a zone or
the dispersal being demanded. Not only is no discussion possible, since the wall of sound is unbreachable, but no protest is
possible in the moment either: it can only be deferred or take
place in another location. These technologies are also totalizing because they break the collective and send everyone back
to their individual selves, to protect themselves or flee. "Why
do people assemble?" asked the very official Institute for Non-
Lethal Defense Technologies. "What determines the intensity
or 'energy' of a gathering?"[64] Thus the collective can surprise,
preoccupy, pose problems. It is a threat to be circumscribed.

Independent researcher Mike Davis, analyzing the emergence of a security-prone urbanism in the 1990s, wrote:

> "Security" has less to do with personal safety than with
> the degree of personal insulation, in residential, work,
> consumption and travel environments, from "unsavory"
> groups and individuals, even crowds in general. . . . More-
> over, the neo-military syntax of contemporary architecture
> insinuates violence and conjures imaginary dangers."[65]

Sound technologies of control play the same role, albeit in a
more violent manner, as "situational prevention" urban fix-
tures and furniture, which aim to separate, contain, orient,
and keep the obligatory flow moving: the machine tolerates
no blockage, no stopping, nothing gratuitous.[66]

Space is thereby unfolding, invisibly, into a succession of
zones, some forbidden and some authorized, and into a collec-
tion of sites where specific and evolving behaviors are banned.

The border is no longer fixed but moves as a function of those it wishes to exclude—and those being excluded can neither anticipate it, since it is furtive, nor protect themselves from it, since ears do not choose what they hear. What García López says of the alarm is also valid for all sound devices, from ultrasonic loudspeakers to the LRAD: the sound border "creates a space of commotion," "a space between parentheses, in anticipation of the gesture that will dissolve the point of inflection."[67] A sanitary and fluctuating parenthesis arises when the undesirable organism rests, where the "disturbance to public order" appears—and it will disappear only when the authority decides to interrupt the alarm or when the undesirable exits the forbidden zone. Sound creates a zone of exclusion that adapts to the geography and its inhabitants, bringing forth a virtual city superposed over the previous city to create a purer version, rid of all chaff.

This moving sound barrier acts on both sides of itself: on those it keeps at bay, but also on those it validates. It "establishes, in the spaces it recomposes, certain ways of expressing oneself, of moving and observing movement, of listening and being listened to, of looking and being looked at: it establishes a sound regime."[68] Thus, "each time an alarm sounds, a sound barrier or any kind of alarm, it creates a sonotopic composition that reproduces the discourse of danger, of insecurity, and of urban disturbance by making it *audible*, by making visible the infraction of the norm and filling our daily activity with public denunciations, glances, and suspicions."[69] Each time that a directional loudspeaker broadcasts an ad or gives an order, it reproduces the discourse of consumption, of domestication and conformity.

The Inferno Intenso forbids passing certain doors, the Mosquito commands you not to loiter, the Audio Spotlight causes a buzz about products on the shelf—all this resounds like the Fortress LA evoked by Mike Davis:

> They set up architectural and semiotic barriers to filter out "undesirables." They enclose the mass that remains, directing its circulation with behaviorist ferocity. It is lured by visual stimuli of all kinds, dulled by musak, sometimes even scented by invisible aromatizers. This Skinnerian orchestration, if well conducted, produces a veritable commercial symphony of swarming, consuming monads moving from one cashpoint to another.[70]

And if some try to contravene this everyday sound regimen, "intruders' pictures are recorded on a computer disk as evidence and city park rangers are alerted. Then loudspeakers warn trespassers that they are being watched and that authorities are on their way."[71] The buildings that make up the city transform into furtive ramparts, half-mechanical, half-biological organisms that survey and sort the human organisms: "The sensory systems of many of Los Angeles's new office towers already include panopticon vision, smell, sensitivity to temperature and humidity, motion detection, and, in a few cases, hearing."[72] And as Steve Goodman concludes, "A sonic architecture of control is also emerging. . . . Vigilant control is no longer merely panoptic but pansensory."[73]

This pansensory control is seen not only at the internal borders of cities but also on the borders between nations. The makers of acoustic technologies are at the forefront here too: EORD, which invented the Israeli Shofar, has for example developed a "virtual fence" system allowing for the "acoustic identification and direction finding of targets" that is usable as an enclosure for prisons, military bases, or countries.[74] Meanwhile, the company CETC International, maker of the Chinese equivalent of the LRAD, has developed "jungle and border surveillance systems."[75] As for Europe, it is working hard to put together high-technology systems "on a continent-wide scale" to protect from outside invasion as well as from internal crowds: in its Seventh Framework Programme for Research, which runs from 2007 to 2013, the European Commission has allotted €1.4 billion to research in matters of security.[76] Among the forty-five projects supported under this framework, five use sound as a weapon of control, such as the Automatic Detection of Abnormal Behaviour and Threats in Crowded Spaces, the Autonomous Maritime Surveillance System, and the Integrated Mobile Security Kit. No doubt the warning devices will quickly find their place in this vast virtual wall.

Steve Wright and Professor Brian Martin analyze this development of "non-lethal" devices at borders as state fears grow about climactic, nuclear, or social catastrophes to come, and the influx of refugees to which they may give rise: "refugee issues are reframed from a matter of humanitarian assistance into a new technopolitics of exclusion."[77] Among the arsenal envisioned, they note anti-personnel Taser mines, lasers causing intense pain, acoustic weapons or vortices, repellant devices that cause nausea or anxiety, and other "bio-regulators," all of which would have the effect of preventing access to the border—and the possibility of exercising one's right of asylum or free circulation. But the wall makers aren't concerned with these details:

> If directed energy weapons are deployed at borders, how could refugees protect themselves? A simple answer was provided by one US official in regard to the Vehicle Mounted Microwave Device which uses microwaves to

heat human skin to unbearable levels: "There are only two sorts of people: tourists and terrorists. Tourists don't use counter measures and we shoot terrorists."[78]

The "sound regime," a temporal and geographic parenthesis within the city's space, is multiplying, branching out, inserting itself in all corners of the urban or geopolitical landscape, to the point of constituting, to borrow Steve Goodman's term, a "politics of frequency."[79] This politics is administered according to its own security-publicity logic of sounds attractive to "tourists" or "consumers" and sounds repellant to "terrorists" or "undesirables," and it is being established de facto, without discussion, a sanitary and private management of public spaces. A willed "reenchantment" of the sound space is occurring, the "Disneylandification of public space" that sociologist Jean-Pierre Garnier speaks about: "The imperative of HQE [high-quality environment], which the crusaders of sustainable urban development fuss about, applies . . . as well to the human environment: only 'quality' people will have the right to frequent the reclassified urban spaces."[80] The sound space is the site of a discreet deployment of dematerialized, arbitrary, and shifting borders. It is becoming an inexhaustible resource that we must, for reasons of security-minded free-market liberalism, make efficient and profitable, privatized, hygienicized.

# CONCLUSION
## "A PASSIONATE SOUND GESTURE"[1]

The recent history of sound as a weapon covers a disparate reality, a mixture of fantasy weapons and real devices. We should neither take comfort in the fantastical nature of the former nor minimize the reality of the latter. The fantasies, first of all, are often residues of prototypes that were imagined but never completed. Or they are the result of rumors propagated by the promoters of acoustic weapons to burnish their image and discourage resistance. In all cases, they serve to make the real devices more worrisome, less easy to grasp, more difficult to attack—even entertaining. The public debate, as a result, is almost in a state of suspension.

Infrasounds remain essentially nonexistent in law enforcement and war, but not for lack of trying: the history of their research is highly inventive and rich. Vortex and plasma weapons will also continue to feed works of fiction first and foremost, even though certain applications are within the realm of possibility. More immediately perilous, however, are "incapacitating" grenades and the use of explosions (detonation cannons or breaking the wall of sound), which today are not subject to any specific regulation. Experimentation in sensory deprivation, silence or saturation, have for their part created a repressive science of human behavior and furnished states with instruments of

torture disguised as entertainment. Medium and high frequencies, finally, are being exploited, whether on the battlefield, as a means of harassment, as a tool of deception, or as
"pre-lethal" weapons. In daily life, they are used to manage
people, borders, no-man's-lands where human material is
sorted.

"The soundscape of the world is changing," wrote composer Murray Schafer in 1977, in analyzing the development
of "noise pollution," the "indiscriminate and imperialistic
spread of more and larger sounds into every corner of man's
life."[2] Less than half a century later, the soundscape is experiencing a new mutation: the spread of sound continues, but it is
now closely controlled, calibrated, fenced off. A form of progress? It is over us, our movements, our imagination that this
control, this calibration, this fencing off aims to apply. To sort
sound in this way is to sort the living. It is to eliminate anything unpredictable, mixed, gratuitous. It is to use sound to
mask reality and to contain it, to ensure nothing goes outside
the approved framework. The pathways to evasion, resistance,
and critical reappropriation are yet to be invented. But the
exploration of sound territories is not the privileged domain
of professionals of repression or lovers of order—far from it.
We will lend an ear here to some "passionate sound gestures"
past and present that have participated in the establishment of
critical imagination on this issue.

What if sound, instead of controlling, surveying, and
separating, was the instrument of the collective, nourished
it, awakened it, amplified it? If it became the tool of a jubi-
lant disorder, of festive mutiny, of reversal? In the early 1970s,
William Burroughs foresaw not only the use of infrasounds
in concerts to make the public vibrate but the broadcasting of
riot sounds to create riots:

I consider the potential of thousands of people with recorders, portable and stationary, messages passed along like signal drums, a parody of the President's speech up and down
balconies, in and out open windows, through walls, over
courtyards, taken up by barking dogs, muttering bums,
music, traffic down windy streets, across parks and soccer
fields. Illusion is a revolutionary weapon. . . . Take a recorded Wallace speech, cut in stammering coughs sneezes
hiccoughs snarls pain screams fear whimperings apoplectic
sputterings slobbering drooling idiot noises sex and animal
sound effects and play it back in the streets subway stations
parks political rallies. . . . Recorded police whistles will
draw cops. Recorded gunshots, and their guns are out.[3]

The entire city, down to its buildings, can even begin to
vibrate under the effect of a "sonic wind." Sound, far from fos-
silizing interior borders and compartmentalizing zones, could
disintegrate stability, and "one building's sound [could] infect
another."[4] American artist Mark Bain is probably not far from
what Burroughs called for when he created his "vibrational
architecture."[5] Installing sensors on buildings or on bridges
that capture their internal vibrations, he amplifies these vibrations, causing the structures to tremble: "The machines
[are] fused to architecture, and [are] playing the building as
an instrument."[6] Bain, an "anti-architect," practices what he
calls an "archeology of technology": he turns the military and
police research on infrasounds on its head, recycling the "scientific debris" to create a "connective tissue" between visitors
and the building, to "question . . . architectural authority," and
to explore the distinction between public and private spaces.[7]
Steve Goodman concludes, "[In] a topology in which every
resonant surface is potentially a host for contagious concepts,

precepts and affects . . . all matter becomes a reservoir of mediatic contagion."[8]

The Spanish collective Escoitar, which is driven by the anthropologist Chiu Longina and musicologist Juan-Gil López, works on uncovering and deciphering the police or military use of sound, but also on developing the critical and creative use of sound via "sound acts." In *Sonic Weapons*, Escoitar tells a history of sonic weapons: it is a sound broadcast in the form of a counterattack, which not only informs about these devices but responds to the security and exclusionary use of some frequencies with a sound composition.[9] The objective of the piece is twofold: "on the one hand to alert the listener to a problem, the use of sound and hearing as a means of control, and on the other hand to produce a passionate sound gesture to defend sounds and to exercise freedom." Escoitar has in addition developed a base of online resources on acoustic weapons and on the effects of sound in general, creates installations and performances, and has created public studios around these technologies. The collective thus proposes to "locate the urban sound devices capable of exercising power and control," to "catalogue and map out these mechanisms on a public map," and to "listen to the city with ears wide open."

Today we must take back awareness of sound, seize it, and counter its co-optation for security and sales, in order to invent uses that allow us to inhabit the common acoustic space differently. If the powers that be mean to invade every aspect of life, we must struggle to ensure that life escapes it—in the realm of sound and elsewhere.

# NOTES

## Introduction: "We Don't Yet Know What a Sonic Body Can Do"

1. Steve Goodman, *Sonic Warfare: Sound, Affect and the Ecology of Fear* (Cambridge, MA: MIT Press, 2010), xvi. A bibliography listing all references and URLs is available on the author's website, www.intempestive .net.

2. This book will build on research published in January 2010 by *Article XI*, an independent website that has also become a printed journal. The initial research is still online at www.article11.info, and other columns have since appeared under the heading "Politiques du son" in the print version. This book also grew out of work with sound on French *radio libres*: Frequence Paris Plurielle at first, then Radio Galère in Marseille.

3. Presentation by the collective Escoitar of its sound play *Sonic Weapons*, produced in 2009 and available at www.artesonoro.org/sonic-weapons/ (accessed April 28, 2011).

4. The effect of a pulsed energy projectile, a "non-lethal" weapon under development.

5. "Cut-up" refers to a literary technique, developed by sound poet Brion Gysin and used by the writer William Burroughs, consisting of cutting up a text and rearranging the fragments in a random fashion to produce another text.

6. Some of these resources are available on the Internet. In an effort to avoid overloading this book with notes, the author has included the links in a bibliography on her website, www.intempestive.net.

7. Neil Davison, *"Non-Lethal" Weapons* (Houndmills, UK: Palgrave Macmillan, 2009).

8. Research reports can be downloaded from the project website, www .brad.ac.uk/acad/nlw (accessed January 23, 2011). The Internet being a flourishing but particularly changeable resource—pages are modified, are moved, and disappear—we have systematically indicated the date on which these websites were consulted.

9. Goodman, *Sonic Warfare*.

## 1. "Ears Don't Have Lids": Technical Aspects of Hearing

1. Pascal Quignard, *La Haine de la musique* (Paris: Calmann-Lévy, 1996), 115ff; Murray Schafer, *The Soundscape: Our Sonic Environment and the Tuning of the World* (Rochester, NY: Destiny Books, 1977, 1994), 11.

2. Michel Chion, *Le Son: Traité d'acoulogie* (Paris: Armand Colin, 2010), 23. Another one of his works is entitled *Le Promeneur écoutant* (Paris: Plume/Sacem, 1993).

3. Sound cannot exist in a vacuum, since there is no matter to propagate it.

4. Chion, *Le Son*, 24–25.

5. Schafer, *Soundscape*, 11.

6. From 1,000 Hz on, frequency is often noted in kilohertz (kHz).

7. The "la" at 440 Hz is a convention, which has had and still has some variations according to different usages.

8. These values are relative: according to the individual and the intensity, infrasounds or ultrasounds can become audible.

9. Antonio Fischetti notes that "some sounds, like the vibration of a tuning fork, a whistle or even a high note emitted by a female voice can approach this" (*Invitation à l'acoustique* [Paris: Belin, 2003], 16).

10. The energy of a flute highlights the basic frequency, while the energy of a trumpet accentuates elevated harmonics (ibid., 25).

11. A subwoofer is a loudspeaker that reproduces very low frequencies, below 80 Hz.

12. Non-acoustic sound waves move in a spherical manner.

13. Jürgen Altmann, "Acoustic Weapons, a Prospective Assessment: Sources, Propagation and Effects of Strong Sound," Occasional Paper no. 22, Cornell University Peace Study Programs, May 1999, 20, www.acoustics.org (accessed December 21, 2010), and also in Michael J. Griffin, *Handbook of Human Vibration* (London: Academic Press, 1990), chap. 7; Henning E. Von Gierke and Donald E. Parker, "Infrasound," in *Handbook of Sensory Physiology*, vol. V/3, *Auditory System: Clinical and Special Topics*, ed. Wolf D. Keidel and William D. Neff (Berlin: Springer-Verlag, 1976). Several versions of the same Altmann study exist: we base ourselves here on the original 1999 version, except when added aspects are relevant to our study, in which case we will indicate the date of publication.

14. Altmann, "Acoustic Weapons," 21. Ultrasounds are based on this principle: the image is obtained thanks to ultrasounds sent to different parts of the body. This is made possible via the difference between the acoustic impedance of air (the relationship of the pressure on the speed of the wave) and that of the body.

15. James R. Jauchem and Michael C. Cook, "High-Intensity Acoustics for Military Non-Lethal Applications: A Lack of Useful Systems," *Military Medicine*, February 2007.

16. Altmann, "Acoustic Weapons," 21.

17. Pierre Schaeffer, *Traité des objets musicaux* (Paris: Seuil, 1966), 97.

18. The decibel is a logarithmic unit and measures the relationship between two powers; thus it is a relative measure. Other measures than the dB

SPL are also used, such as dB(A) and dB(B), which are based on weighting and take into account the particular sensitivity of the human ear, or dBV, which measures tension relative to a reference of 1 volt.

19. Amplification is the raising of the amplitude of sound, whether it occurs naturally (by raising one's voice, placing one's hands around one's mouth), architecturally (the acoustics of Greek amphitheaters), or electronically (through a computer, amplifier, or loudspeakers).

20. Boris Jollivet, in particular, provides the sounds of the wolf spider on the CD *Orchestre animal*, on the label Chiff Chaff. David Dun has recorded the sound of beetles in pine trees and turned it into *The Sound of Light in Trees*, on the EarthEar label.

21. The permanent threshold shift, which in general affects the frequency of 4,000 Hz, then spreads to lower and higher frequencies. See Altmann, "Acoustic Weapons," 12.

22. Ibid.

23. Ibid., 13.

24. It is also the number indicated by various scientists cited in Altmann, "Acoustic Weapons," or in Nick Broner, "The Effects of Low Frequency Noise on People: A Review," *Journal of Sound and Vibration* 58, no. 4 (1978), reprinted in *Amok Journal: A Compendium of Psycho-Physiological Investigations*, ed. Stuart Swezey (Los Angeles: Amok Books, 1995), 414.

25. Directive 2003/10/EC of the European Parliament and of the Council of February 6, 2003, on the minimum health and safety requirements regarding the exposure of workers to the risks arising from physical agents (noise), 2003.

26. Tinnitus is a buzzing or ringing sound in the ear that is not emitted by an external source.

27. Altmann, "Acoustic Weapons," 13, 19, 20, 31–35, 55; Neil Davison, *"Non-Lethal" Weapons* (Houndmills, UK: Palgrave Macmillan, 2009), 187.

28. Haiku by Bashō, trans. Michael R. Burch, www.thehypertexts.com /Best%20Haiku.htm (accessed July 11, 2012).

29. Antonio Fischetti, *Initiation à l'acoustique* (Paris: Belin, 2003), 107.

30. Certain animals, such as horses or dogs, have ear flaps that can be aimed, which allows them to detect the origin of a sound more precisely and faster.

31. Joachim Ernst-Berendt, *The Third Ear* (Perth: Element, 1985), 79, quoted in Steve Goodman, *Sonic Warfare: Sound, Affect and the Ecology of Fear* (Cambridge, MA: MIT Press, 2010), 65.

## 2. The Death Ray: Infrasounds and Low Frequencies

1. Pierre Liénard, *Petite histoire de l'acoustique: Bruits, sons et musique* (Paris: Lavoisier, 2002), 356.

2. Reported in Donald Tuzin, "Miraculous Voices: The Auditory Experience of Numinous Objects," *Current Anthropology* 25, no. 5 (December 1984), reprinted in *Amok Journal: A Compendium of Psycho-Physiological Investigations*, ed. Stuart Swezey (Los Angeles: Amok Books, 1995), 430.

3. Jürgen Altmann, "Acoustic Weapons, a Prospective Assessment: Sources, Propagation and Effects of Strong Sound," Occasional Paper no. 22, Cornell University Peace Study Programs, May 1999, 16.

4. Geoff Leventhall, Peter Pelmear, and Stephen Benton, "A Review of Published Research on Low Frequency Noise and Its Effects," DEFRA, May 2003, 55–56.

5. Altmann, "Acoustic Weapons," 19.

6. Ibid., 18.

7. Nick Broner, "The Effects of Low Frequency Noise on People: A Review," *Journal of Sound and Vibration* 58, no. 4 (1978), reprinted in *Amok Journal*, 410. Many authors refuse to treat infrasounds and low frequencies separately.

8. William Arkin, "Acoustic Anti-Personnel Weapons: An Inhuman Future?" *Medicine, Conflict and Survival* 14 (1997): 316.

9. Dr. Michael Bryan and Dr. William Tempest, "Does Infrasound Make Drivers 'Drunk'?" *New Scientist*, March 16, 1972, reprinted in Swezey, *Amok Journal*, 396.

10. Vladimir Gavreau, "Infrasound," *Science Journal* 4, no. 1 (January 1968), reprinted in Swezey, *Amok Journal*, 382.

11. Many of these sources are Anglo-Saxon, such as Daria Vaisman, "The Acoustics of War," *Cabinet*, no. 5 (winter 2001–2).

12. Liénard, *Petite histoire de l'acoustique*, 354.

13. Gavreau, "Infrasound," 379.

14. Steve Goodman, *Sonic Warfare: Sound, Affect and the Ecology of Fear* (Cambridge, MA: MIT Press, 2010), 18.

15. Jacques Lesinge, interview with Vladimir Gavreau and Albert Calaora, *L'Aurore*, May 30 1967, quoted in Liénard, *Petite histoire de l'acoustique*, 354.

16. Gavreau, "Infrasound," 379.

17. Lesinge, interview with Gavreau and Calaora.

18. Swezey, *Amok Journal*, 372.

19. Liénard, *Petite histoire de l'acoustique*, 355.

20. Gavreau, "Infrasound," 386.

21. Ibid., 382.

22. Robert Levavasseur, a colleague of Gavreau's, built police whistles amplified by a vibrating chamber, which thereby emitted various frequencies (340 Hz, 2,600 Hz) at sonic levels up to four hundred times stronger than those of an ordinary police whistle (Gavreau, "Infrasound," 381).

23. Ibid., 382.

24. Ibid., 383.

25. Liénard, *Petite histoire de l'acoustique*, 355.

26. Jürgen Altmann, "Acoustic Weapons: A Prospective Assessment," in *Science and Global Security* 9, no. 3 (2001): 227; interview with George Canevet of the Laboratoire de Mécanique et d'Acoustique (LMA) in Marseille.

27. William S. Burroughs, "The Jimmy and Bill Show," *Crawdaddy*, 1973, in Swezey, *Amok Journal*, 376.

28. Altmann, "Acoustic Weapons" (1999), 35.

29. Ibid., 18–20. The scars were observed among World War II German submariners and mentioned in Daniel Johnson, "The Effects of High Level Infrasound," in *Conference on Low Frequency Noise and Hearing, 7–9 May 1980 in Aalborg, Denmark*, ed. Henrik Moller and Per Rubak (Aalborg: Aalborg University Press, 1980).

30. Stanley Harris and Daniel Johnson, "Effects of Infrasound on Cognitive Performance," *Aviation, Space and Environmental Medicine* 49, no. 4 (April 1978), quoted in Michael Cook, Clifford Sherry, Caroll Brown, et al., "Lack of Effects on Goal-Directed Behavior of High-Intensity Infrasound in a Resonant Reverberent Chamber," U.S. Air Force Research Laboratory, November 2001, 31.

31. Hendricus G. Loos, U.S. patent 6017302: "Subliminal Acoustic Manipulation of Nervous Systems," 2000. The author claims that "the effects of the 2.5 Hz resonance include slowing of certain cortical processes, sleepiness and disorientation."

32. Broner, "Effects of Low Frequency Noise," 419.

33. Promoters of acoustic weapons quoted in Arkin, "Acoustic Anti-Personnel Weapons," 316.

34. Broner, "Effects of Low Frequency Noise," 419.

35. Ibid.

36. Rex Applegate, *Riot Control: Material and Techniques* (Harrisburg, PA: Stackpole, 1969), 273.

37. "Army Tests New Riot Weapon," *New Scientist*, September 20, 1973, quoted in Swezey, *Amok Journal*, 400.

38. "Anti-Crowd Weapons Work by Causing Fits," *New Scientist*, March 29, 1973, in Swezey, *Amok Journal*, 402–3; Carol Ackroyd, Karen Margolis, Jonathan Rosenhead, and Tim Shallice, *The Technology of Political Control* (Harmondsworth, UK: Penguin Books, 1977), 225–26.

39. Neil Davison, *"Non-Lethal" Weapons* (Houndmills, UK: Palgrave Macmillan, 2009), 189.

40. Research aiming to exploit the superposition of two ultrasonic frequencies to produce a third (not necessarily infrasonic) were financed in the 1960s by the Office of Naval Research, but with an entirely different objective: physicist Peter Westervelt developed a parametric array, which made it possible to perfect more directional sonars that would find uses in the medical field. See Peter J. Westervelt, "Parametric Acoustic Array," *Journal of the Acoustical Society of America* 35, no. 4 (April 1963): 535–37.

41. Altmann, "Acoustic Weapons" (1999), 47–48.

42. Hungary, "Working Paper on Infra Sound Weapons," Report to the United Nations: CCD/575, August 14, 1978, in Swezey, *Amok Journal*, 405–9.

43. M.T., "Russians Continue Work on Sophisticated Acoustic Weaponry," *Defense Electronics*, no. 26 (March 1994).

44. Altmann, "Acoustic Weapons" (1999), 48–51.

45. Applied Research Laboratories, "Non-Lethal Swimmer Neutralization Study," University of Texas, Austin, G2 Software Systems Inc., May 2002, 23, available at www.spawar.navy.mil (accessed on December 21, 2010).

46. "New Concept Weapon and Its Medical-Related Problems," *Beijing Renmi Junyi*, no. 9 (1997): 507–8, quoted in James R. Jauchem and Michael C. Cook, "High-Intensity Acoustics for Military Non-Lethal Applications: A Lack of Useful Systems," *Military Medicine*, February 2007.

47. Arkin, "Acoustic Anti-Personnel Weapons," 320.

48. Ibid., 321; Davison, *"Non-Lethal" Weapons*, 190.

49. SARA, "Selective Area/Facility Denial Using High Power Acoustic Beam Technology," *Phase 1 SBIR Final Report*, March 10, 1995 (revised February 13, 1996), cited in Arkin, "Acoustic Anti-Personnel Weapons," 316.

50. Arkin, "Acoustic Anti-Personnel Weapons," 316.

51. Ibid., 322; Davison, *"Non-Lethal" Weapons*, 190–91; ARDEC, "ARDEC Exploring Less-than-Lethal Munitions to Give Army Greater Flexibility in Future Conflicts," news release, October 9, 1992, quoted in Davison, *"Non-Lethal" Weapons*, 190; David Habling, *Weapons Grade: The Revealing History of the Link Between Modern Warfare and Our High Tech World* (London: Constable & Robinson, 2005), 242; "Law Enforcement Applications (Non-Lethals)," section of the SARA website in 2006, web. archive.org/web/20061116042950/http://www.sara.com (accessed December 23, 2010).

52. Jürgen Altmann quoted in Davison, *"Non-Lethal" Weapons*, 191; see also Altmann, "Acoustic Weapons" (1999), 4–5.

53. Arkin, "Acoustic Anti-Personnel Weapons," 322.

54. Davison, *"Non-Lethal" Weapons*, 191.

55. Ibid., 192.

56. We have chosen to discuss SADAG in this chapter rather than in the one on explosions, given that the objective of AFRL's research was to study the effects of infrasounds and not explosions in and of themselves.

57. Clifford Sherry, Michael Cook, Carroll Brown, et al., "An Assessment of the Effects of Four Acoustic Energy Devices on Animal Behavior," AFRL, October 2000, ii.

58. The sirens and the Gayl blaster did not emit infrasonic frequencies but medium or high frequencies.

59. Sherry et al., "Assessment of the Effects," 67.

60. Definition of "non-lethal" weapons by the U.S. Department of Defense, "Policy for Non-Lethal Weapons," Directive 3000.3, art. 3, July 9, 1996.

61. U.S. Department of Justice, Office of Justice Programs, National Institute of Justice, "NIJ Awards in Fiscal Year 1997," June 1998.

62. Michael Murphy, James Jauchem, and Michael Merrit, "Acoustic Bioeffects Research for Non-Lethal Applications," Proceedings of the First European Symposium on Non-Lethal Weapons, Ettlingen, Germany, September 25–26, 2001, quoted in Davison, *"Non-Lethal" Weapons*, 193.

63. Cook et al., "Lack of Effects," 31.

64. Ibid., 32.

65. Department of Defense, SBIR FY97.2, "Solicitation Selections w/ Abstracts. Army. 199 Phase I Selections from the 97.2 Solicitation. Synetics

Corp: Parametric Difference Waves for Low Frequency Acoustic Propagation," 1997.

66. Synetics website in 2001, web.archive.org/web/20010913155048/www.synetics.com/ (accessed December 20, 2010).

67. Joint Non-Lethal Weapons Program, 1999 Annual Report, 19.

68. G. Alker, "Acoustic Weapons: A Feasibility Study," Defense Evaluation and Research Agency, April 1996, quoted in Sherry et al., "Assessment of the Effects," 68.

69. Jauchem and Cook, "High-Intensity Acoustics."

70. Davison, *"Non-Lethal" Weapons*, 200.

71. "The Army's Center for Lethality," ARDEC's slogan on its website, www.pica.army.mil/picatinnypublic/organizations/ardec/index.asp (accessed January 24, 2011).

72. "Infrasonic Stimuli" section of the SMBI website, njms2.umdnj.edu/smbiweb/index.html (accessed February 26, 2011).

73. Davison, *"Non-Lethal" Weapons*, 200.

74. Applied Research Laboratories, "Non-Lethal Swimmer Neutralization Study," 20. This is due to the fact that the acoustic blocking effect of the water and that of the human body (made up mostly of water) are very close.

75. Ibid.

76. Ibid., 39–40, 71.

77. Ibid., iii, 47.

78. International law stipulates that weapons must be discriminating, that they must enable the distinction between a combatant and a noncombatant (civilians, the wounded, journalists, etc.).

79. This barrier is being used commercially by the Belgian company ProFish Technology.

80. Technical specifications for the Rumbler on the manufacturer's website, www.fedsig.com (accessed January 30, 2011).

81. "The Rumbler," *Weird Vibrations*, November 2009, www.weird vibrations.com (accessed January 30, 2011).

82. Title of a study by Vic Tandy and Tony R. Lawrence, "The Ghost in the Machine," *Journal of the Society for Psychical Research* 62, no. 851 (April 1998).

83. Ibid.

84. Vic Tandy, "Something in the Cellar," *Journal of the Society for Psychical Research* 64, no. 3 (July 2000).

85. The bullroarer is a small plank attached to a cord that you swing in the air to produce sound.

86. Tuzin, "Miraculous Voices," 430.

87. Ibid., 432.

88. Ibid., 433.

89. Ibid., 430.

90. The experiment is described on the website of Sarah Angliss, www.saraangliss.com (accessed January 22, 2011) and by Jonathan Amos, "Organ Music 'Instills Religious Feelings,'" BBC News, September 8, 2003.

91. Swezey, *Amok Journal*, 373–74.

92. Hannah Magill, "Gaspar Noé," British Film Institute, interview at the National Film Theatre, October 11, 2002.

93. Sarah Angliss, "Postscript: Deep Weirdness or Hot Air?" www .sarahangliss.com/infrasonic-postscript (January 30, 2011).

## 3. "Hit by a Wall of Air": Explosions

1. Palestinian testimony in Chris McGreal, "Palestinians Hit by Sonic Boom Air Raids," *The Guardian*, November 3, 2005.

2. Curtis Roads, *Microsound* (Cambridge, MA: MIT Press, 2001), 7, quoted by Steve Goodman, *Sonic Warfare: Sound, Affect and the Ecology of Fear* (Cambridge, MA: MIT Press, 2010), 10.

3. Jürgen Altmann, "Acoustic Weapons, a Prospective Assessment: Sources, Propagation and Effects of Strong Sound," Occasional Paper no. 22, Cornell University Peace Study Programs, May 1999, 19; Applied Research Laboratories, "Non-Lethal Swimmer Neutralization Study," University of Texas, Austin, G2 Software Systems Inc., May 2002, 39.

4. Altmann, "Acoustic Weapons," 29; Applied Research Laboratories, "Non-Lethal Swimmer Neutralization Study," 19.

5. Altmann, "Acoustic Weapons," 31–35.

6. Applied Research Laboratories, "Non-Lethal Swimmer Neutralization Study," 40–41.

7. Allegation of SARA on its website in 2006, web.archive.org/web /20061116042950/http://www.sara.com (accessed December 23, 2010), regarding sensations produced by the vortex launcher.

8. The book was published by Wiley & Sons in 1947, then by Chapman & Hall in 1948, and was reprinted by WE in 1971 under the title *Secret Weapons of the Third Reich: German Research in World War II*. We are basing ourselves on this last edition. It should be noted that the observations and photographs brought back by Simon have been regularly reprinted on the Internet, but the source is rarely cited and the information is frequently distorted (whether with regard to the scientists involved, the spelling of their names, the research centers where they worked, or the weapons themselves).

9. Major General Leslie E. Simon, *Secret Weapons of the Third Reich: German Research in World War II* (Old Greenwich, CT: WE, 1971), 183.

10. We see this, for example, in tornados, when water drains from a tub, or when someone makes "rings" with cigarette smoke.

11. Ibid., 184.

12. Ibid., 180.

13. Ibid., 181.

14. George Lucey and Louis Jasper, "Vortex Ring Generator," U.S. Army Research Laboratory, DE Effects and Mitigation Branch (n.d.).

15. David Hambling, *Weapons Grade: How Modern Warfare Gave Birth to Our High-Tech World* (New York: Carroll & Graf, 2005), 250.

16. Neil Davison, *"Non-Lethal" Weapons* (Houndmills, UK: Palgrave Macmillan, 2009), 190.

17. Lucey and Jasper, "Vortex Ring Generator."

18. Ibid.

19. According to the United Nations definition, an incapacitating agent is a chemical weapon, "which produces temporary disabling conditions which can be physical or mental and persist for hours or days after exposure to the agent has ceased."

20. JNLWD, "1997. A Year in Review, Joint Non Lethal Weapons Program," annual report, 1998.

21. National Research Council (2003), quoted by Davison, *"Non-Lethal" Weapons*, 194.

22. Neil Davison and Nick Lewer, "Research Report No. 4," BNLWRP, December 2003, 9.

23. Neil Davison, *"Non-Lethal" Weapons*, 201.

24. A Palestinian witness quoted in McGreal, "Palestinians Hit."

25. Simon, *Secret Weapons*, 181–83.

26. U.S. Senate Committee on Human Resources, "Project MKUltra, the CIA's Program of Research in Behavioral Modification," 95th Cong., 1st sess., August 3, 1977, available at www.nytimes.com/packages/pdf/national/13 inmate_ProjectMKULTRA.pdf (accessed November 30, 2010), 40.

27. Ibid., 165, 168.

28. Ibid., 40.

29. Justin Mullins, "Voice of God," *New Scientist*, December 25, 1999.

30. H. Edwin Boesch, Bruce Benwell, and Vincent Ellis, "High Power Electrically Driven Impulsive Acoustic Source for Target Experiments and Area-Denial Applications," ARL, 1998.

31. Applied Research Laboratories, "Non-Lethal Swimmer Neutralization Study," 43.

32. Ibid., 44.

33. Ibid.

34. Applied Research Laboratories, "Non-Lethal Swimmer Neutralization Study." The Defense Nuclear Agency became the Defense Special Weapons Agency in 1996 and was absorbed into the Defense Threat Reduction Agency in 1998.

35. Applied Research Laboratories, "Non-Lethal Swimmer Neutralization Study."

36. Henry Sze, Charles Gilman, Jim Lyon, et al., "Non-Lethal Weapons: An Acoustic Blaster Demonstration Program," Non-Lethal Defense III, Primex Physics International Company, February 1998.

37. David Hambling, "Plasma Shield May Stun and Disorientate Enemies," *New Scientist*, April 26, 2007.

38. Stellar Photonics website, www.sphotonics.com/Programs-PASS (accessed February 10, 2011).

39. Keith Braun, of the Picatinny Arsenal, quoted in Hambling, "Plasma Shield."

40. David Hambling, "Plasma Lasers for Shielding—and Advertising," *Wired*, May 4, 2009. All articles cited from *Wired* were consulted on the magazine's website, www.wired.com, mainly under the column "Danger Room," between the months of November 2010 and May 2011.

41. See Neil Davison, "The Early History of 'Non-Lethal' Weapons," BNLWRP, December 2006, 4.

42. Ministry of Agriculture, Food and Rural Affairs, Canada, "Using Propane-Fired Cannons to Keep Birds Away From Vineyards," *Agdex* 730, no. 658 (July 2010).

43. Named after pulse detonation technology (PDT), which uses liquefied natural gas as a combustible.

44. The two products are visible on Armytec's website, www.armytec .net (accessed December 23, 2010).

45. Barbara Opall-Rome, "A Cannon Stun Gun," *Defense News*, January 11, 2010.

46. Ibid.

47. Ibid.

48. Ibid.

49. There are "air cannons" that produce a higher amplitude (up to 230 dB in the water, or 255 dB if several cannons are arrayed), but they are used in seismic prospecting, to locate deposits of hydrocarbons. The shock waves generated on these occasions are not without impact on marine life, as attested by a report of the Union québécoise pour la conservation de la nature, "Les impacts environnementaux de l'exploration pétrolière et gazière dans le golfe du Saint-Laurent," December 2003, 10ff.

50. Presentation of the shock wave cannon on the Armytec website.

51. Gideon Levy, "Demons in the Skies of the Gaza Strip," *Haaretz*, November 6, 2005.

52. Ibid.

53. McGreal, "Palestinians Hit."

54. Ibid.

55. Anat Bershkovsky, "Living with Supersonic Booms," Ynetnews, October 30, 2005.

56. McGreal, "Palestinians Hit."

57. Ibid.

58. Ibid.

59. Yuval Yoaz, "State Tells Court: 'Sonic Booms over Gaza to Confuse Terrorists,'" *Haaretz*, November 14, 2005.

60. McGreal, "Palestinians Hit."

61. Physicians for Human Rights, "Report: Harm to Children in Gaza, November 2006," November 8, 2006, 2.

62. Human Rights Council, "Report of the Special Rapporteur on the Situation of Human Rights in the Palestinian Territories Occupied since 1967 (John Dugard)," A/HRC/7/17, January 21, 2008, 5, 8.

63. United Nations, "UN Human Rights Expert Welcomes Landing of Relief Vessels in Gaza," news release, August 25, 2008.

64. Yoav Stern, "Report: Amid Gaza Op, IAF Sets Off Sonic Booms over Lebanon," *Haaretz*, December 28, 2008.

65. World Bank, "Fact Sheet: Gaza Strip Water and Sanitation Situation," January 7, 2009, go.worldbank.org (accessed February 17, 2010). The repair of the purification station, undertaken within the framework of the North Gaza Emergency Sewage Treatment Project, produced its first results in January 2010.

66. Presentation of the Multi-Sensory Grenade by SARA on its website in 2006.

67. Rex Applegate, *Riot Control: Material and Techniques* (Harrisburg, PA: Stackpole, 1969), 301.

68. Alsetex, "Brevet d'invention pour grenade explosive sans éclats," patent filed in France, August 7, 1970 (no. 70.29341) and in Belgium July 1, 1971 (no. 769379).

69. Ibid.

70. Malvern Lumsden, *Anti-Personnel Weapons* (London: Taylor & Francis, 1978), 210, quoted in Davison, "Early History," 19.

71. The candela (cd) is a unit of measure of luminous intensity. A candle has an intensity of 1 cd, a 100-watt filament lamp has approximately 110 cd, and a 125-watt halogen lamp has approximately 200 cd.

72. Peter Padley, "Grenades of the Army," www.hmforces.co.uk, June 14, 2010 (accessed February 11, 2011).

73. Neil Davison, "The Contemporary Development of 'Non-Lethal' Weapons," BNLWRP, May 2007, 32.

74. Neil Davison, "The Development of 'Non-Lethal' Weapons During the 1990's," BNLWRP, March 2007, 12.

75. Sandia National Laboratories, Law Enforcement Technologies, Inc., and Martin Electronics, Inc., "Variable Range Less-than-Lethal Ballistic, Final Report," 2002, Department of Justice, National Institute of Justice, quoted in Davison, "Contemporary Development," 32.

76. Davison, "Contemporary Development."

77. U.S. Marine Corps, "Sources Sought Notice: 10—PM Force Protection Systems (FPS) Clear-A-Space Distract/Disorient Program Request (Ref.: M67854-FPS-RFI-04-0001)," *Federal Business Opportunities Daily*, June 15, 2004.

78. Presentation of the Multi-Sensory Grenade by SARA on its website in 2006.

79. U.S. Army ARDEC, "Sources Sought Notice: 13. M84 and XM012 Hand Grenade (Stun) (Ref.: W15QKN-04-X-0101)," *Federal Business Opportunities Daily*, November 13, 2003.

80. Neil Davison and Nick Lewer, "Research Report No. 6," BNLRWP, October 2004, 40–41.

81. E-Labs, Inc., "Performance Characterization Study: Noise Flash Diversionary Devices (NFDDs)," June 1, 2004, ii, 15. The test was of grenades made by ALS Technologies, Combined Tactical Systems, Defense Technologies, NICO Pyrotechnics, Precision Ordnance, and Pyrotechnic Specialties. The company Non-Lethal Technologies should also be noted; it furnishes flash-bang grenades that have an advertised sonic intensity of 175 dB and a luminosity of 7 million cds, www.non-lethaltechnologies.com/DD-FB.htm (accessed February 12, 2011).

82. Sandia, "Sandia Licenses Its Improved Flash-Bang Technology," news release, April 15, 2008.

83. Vince Beiser, "Flash and Awe: A Better Stun Grenade Protects the Good Guys," *Wired*, September 22, 2008.

84. Department of Homeland Security, U.S. Coast Guard, "Non-Pyrotechnic Flash-Bang Grenade (NPFG)," *Federal Business Opportunities Daily*, March 16, 2010.

85. Cleve R. Wootson Jr., "Police: Error Led to Veteran SWAT Officer's Death," *Charlotte Observer*, March 17, 2011.

86. Interpolitex, "Rosoboronexport at Interpolitex 2005," October 10, 2005.

87. See the company's online catalog at www.rheinmetall.de (accessed February 3, 2011).

88. Palestinian Center for Human Rights, "Weekly Report on Israeli Human Rights Violations in the Occupied Palestinian Territory (May 5–11, 2011)," May 12, 2011; Reuters, "La mort de manifestants avive les tensions en Albanie," *The Tribune*, January 22, 2011; SAPA, "Police Fire Stun Grenade at World Cup Guard Protest," *Mail & Guardian*, June 17, 2010; Sarah A. Topol, "Égypte: comment mobiliser 2 millions de personnes sans Internet ni SMS en une journée," *Slate*, February 2, 2011; "Affrontements sanglants au Kirghizistan," *Le Monde*, April 7, 2010; Administrative and Financial Division of the Columbian National Police, "Adicional en plazo no 3 al contrato de compraventa PN Diraf no 06-2-10574-07 Celebrado entre la Policía nacional y la firma importadora y distribuidora de Colombia LTDA. Imdicol apoderado en Colombia de la firma Combined systems inc.," March 30, 2010.

89. Lacroix-Ruggieri website, www.lacroix-ruggieri.com (accessed February 13, 2011).

90. Including the manual protection devices, which we discuss further along. See Lacroix's online catalog of defense and security items, www.lacroix ds.com/catalogue.html (accessed February 12, 2011).

91. Alsetex website in 2005, web.archive.org/web/20051012155832 /www.alsetex.fr (accessed February 14, 2011).

92. Alexandre Targé, "Les armes non létales: l'exemple de Davey Bickford," *Artem Défense* (newsletter), January 2008.

93. A more complete article on DMPs was published in *Jusqu'ici*, November 6, 2010, jusquici.toile-libre.org (accessed February 8, 2011).

94. Information sheet for DBD95/DMP, www.sapl-sas.com/fr/produits /index/id:19/ (accessed February 11, 2011).

95. Jacky Durand, "Gare à l'orthochlorobenzylidénémalononitrile," *Libération*, March 23, 2006; testimony of another retiree from Grenoble in Commission nationale de déontologie de la sécurité (CNDS), "Étude sur l'usage des matériels de contrainte et de défense par les forces de l'ordre," annual reports, 2009.

96. Durand, "Gare."

97. Jean Delavaud, "Pascal, quarante-deux ans, blessé par une grenade," *Ouest France*, February 17, 2009.

98. "L'étudiante avait perdu un œil: trois policiers mis en examen," *Le Dauphiné libéré*, December 16, 2008; Mario, "L'après-manif de Grenoble contre les nanotechnologies," *Le Numéro zéro*, June 9, 2006, lenumerozero .lautre.net (accessed May 15, 2011); Hélène Jeanmougin, "Contre-sommet de l'OTAN, Strasbourg 03/04/2009," *Bon pied bon œil*, April 3, 2009, bonpiedbonoeil.wordpress.com (accessed February 12, 2011).

99. CNDS, "Étude sur l'usage des matériels."

100. "Heurts à Lorient. Un enseignant porte plainte," *Le Télégramme*, October 22, 2010; Nancy Ladde, "Seize gendarmes blessés à l'entraînement," *Sud-Ouest*, March 2, 2010.

101. UN Protocol on Blinding Laser Weapons (Protocol IV to the 1980 Convention), Geneva, October 13, 1995.

102. UN Convention on Prohibitions or Restrictions on the Use of Certain Conventional Weapons Which May Be Deemed to Be Excessively Injurious or to Have Indiscriminate Effects, Geneva, October 10, 1980.

## 4. "Totally Cut Off from the Known": Silence and Saturation

1. Excerpt from the CIA's *KUBARK Counterintelligence Interrogation*, July 1963, www.gwu.edu (accessed February 25, 2011) defining interrogation techniques via sensory deprivation.

2. John Marks, *The Search for the "Manchurian Candidate": The CIA and Mind Control: The Secret History of the Behavioral Sciences* (New York: W.W. Norton, 1991), 23.

3. "Depuis l'intérieur," *L'Envolée*, January 2002.

4. The book was initially published by Times Books. We are quoting from the reissue, published by W.W. Norton in 1991.

5. Alfred W. McCoy, *A Question of Torture: CIA Interrogation from the Cold War to the War on Terror* (New York: Metropolitan, 2006), 25.

6. Marks, *Search*, 138.

7. Lawrence E. Hinkle Jr. and Harold G. Wolff, "Communist Interrogation and Indoctrination of 'Enemies of the State': Analysis of Methods Used by the Communist State Police (A Special Report)," *Archives of Neurology and Psychiatry*, no. 76 (1956), quoted in McCoy, *Question of Torture*, 45–46; Marks, *Search*, 136–38.

8. The director of the medical unit of the CIA in 1952, quoted in McCoy, *Question of Torture*, 23.

9. Ibid., 25.

10. Marks, *Search*, 23–36; McCoy, *Question of Torture*, 26–27.

11. McCoy, *Question of Torture*, 28–29, 33–34.

12. Ibid., 31.

13. Marks, *Search*, 151.

14. Excerpt from a letter by Ulrike Meinhof, member of the Red Army Faction, in 1972, when she was placed in isolation in the soundproof cell in the Cologne-Ossendorf prison in Germany; French translation available on the Marxiste-Leniniste-Maoïst website, www.contre-informations.fr/doc-inter /allemagne/newallem3.html (accessed April 3, 2011).

15. McCoy, *Question of Torture*, 32, 35.

16. Donald O. Hebb, E.S. Heath, and E.A. Stuart, "Experimental Deafness," *Canadian Journal of Psychology* 8, no. 3 (1954). All quotes in this paragraph are taken from this article.

17. W.H. Bexton, W. Heron, and T.H. Scott, "Effects of Decreased Variation in the Sensory Environment," *Canadian Journal of Psychology* 8, no. 2 (1954). All quotes in this paragraph are taken from this article.

18. McCoy, *Question of Torture*, 35.

19. John C. Lilly, "Mental Effects of Reduction of Ordinary Levels of Physical Stimuli on Intact, Healthy Persons," *Psychiatric Research Reports* 5, no. 1 (1956), quoted in McCoy, *Question of Torture*, 39. Lilly envisaged sensory deprivation not as an instrument of torture but as a moment of relaxation and exploration of the frontiers of consciousness. He later published a work compiling his research: *The Deep Self: Profound Relaxation and the Isolation Tank Technique* (New York: Simon & Schuster, 1977).

20. Marks, *Search*, 152.

21. Patients in psychiatric hospitals, alleged double agents, prisoners, or even, for drug experiments, children at a summer camp or the clients of prostitutes. See McCoy, *Question of Torture*, 29, 44.

22. Ibid., p. 50.

23. CIA, *KUBARK Counterintelligence Interrogation*.

24. This technique, also called "stress postures," constrains the detainee to hold uncomfortable positions for hours (arms outstretched, squatting, hands attached to a rail on the ceiling, standing on tiptoes) so that the body is in a state of permanent tension and imbalance and the detainee cannot rest or sleep.

25. Ibid., 86–87.

26. Lilly, "Mental Effects," quoted in CIA, *KUBARK Counterintelligence Interrogation*, 88.

27. Ibid., 90.

28. So named by the CIA because they end only with the death of the guinea pig or his psychological destruction.

29. McCoy, *Question of Torture*, 60, 64, 65, 71, 88.

30. This manual was written to assist training of Honduran officers and includes passages from the *KUBARK Manual* almost verbatim. McCoy, *Question of Torture*, 86, 92–94.

31. Florence Hervey, ed., *Un monde tortionnaire* (Paris: ACAT-France, 2010), 278.

32. Anne Steiner and Loïc Debray, *RAF: Guérilla urbaine en Europe occidentale* (Montreuil: L'Échappée, 2006), 42. The history of this research is developed in "Nouveaux perfectionnements scientifiques des techniciens de torture," in *À propos du procès Baader-Meinhof, fraction armee rouge: De la torture dans les prisons de RFA*, ed. Klaus Croissant (Paris: Bourgois, 1975), 69–81.

33. A hallucination in which one sees one's own body or body parts.

34. Sjef Teuns, "La torture par privation sensorielle," in Croissant, *À propos du procès Baader-Meinhof*, 65–66.

35. Members of the Movement 2 June and of the Socialist Patients' Collective (in German Sozialistisches Patientenkollektiv, also known as the SPK) were also subject to this treatment. The detainees were placed in isolation and regularly frisked, their visitors limited, and their mail censored.

36. Steiner and Debray, *RAF*, 39, 40, 44.

37. Ibid., 40.

38. Excerpts from this letter are published in Croissant, *À propos du procès Baader-Meinhof*, 108–10. The author has worked from the translation "Lettre du couloir de la mort. 1972," published on the Marxiste-Leniniste-Maoïst website, www.contre-informations.fr/doc-inter/allemagne/newallem3.html (accessed April 3, 2011).

39. Steiner and Debray, *RAF*, 40.

40. Ibid., 42, 54.

41. Ibid., 45.

42. Comité contre la torture des prisonniers politiques en RFA, "La section silencieuse, forme la plus dure de la torture par l'isolement," in Croissant, *À propos du procès Baader-Meinhof*, 88.

43. Teuns, "La torture par privation sensorielle," 63–64.

44. An invention of Leonard Rubinstein in the context of research on sensory deprivation for the project MKUltra, quoted by Marks, *Search*, 145.

45. McCoy, *Question of Torture*, 37–38.

46. Ibid., p. 43.

47. Marks, *Search*, 147.

48. McCoy, *Question of Torture*, 44.

49. Marks, *Search*, 145–46.

50. McCoy, *Question of Torture*, 44.

51. Marks, *Search*, 144.

52. Ibid., 144, 149, 150.

53. Alfred McCoy, *Question of Torture*, 44.

54. Ibid., 45.

55. S. Smith and W. Lewty, "Perceptual Isolation in a Silent Room," *The Lancet*, September 12, 1959, quoted in McCoy, *Question of Torture*, 3.

56. S. Smith, J.C. Kenna, and G.F. Reed, "Effects of Sensory Deprivation," *Proceedings of the Royal Society of Medicine* 55 (December 1962): 1003–14.

57. McCoy, *Question of Torture*, 53–54.

58. Ibid., 54.

59. "Report of the Committee of Privy Counsellors Appointed to Consider Authorised Procedures for the Interrogation of Persons Suspected of Terrorism" (Parker Report), The Majority Report, 3, quoted in European Commission, "Application no 5310/71. Ireland v. the United Kingdom. Report of the Commission," January 25, 1976, 393.

60. Ibid., sec. III-B.

61. Carol Ackroyd, Karen Margolis, Jonathan Rosenhead, and Tim Shallice, *The Technology of Political Control* (Harmondsworth, UK: Penguin, 1977), 34.

62. European Court of Human Rights, "Case of Ireland v. the United Kingdom (Application No. 5310/71), Judgment," Strasburg, January 18, 1978.

63. European Commission, "Application no 5310/71," 247.

64. Ackroyd et al., *Technology of Political Control*, 33.

65. European Commission, "Application no 5310/71," 402.

66. Ibid.

67. Ackroyd et al., *Technology of Political Control*, 35.

68. European Commission, "Application no 5310/71"; McCoy, *Question of Torture*, 57–58.

69. European Court of Human Rights, "Case of Ireland v. the United Kingdom," sec. III-B, C, G.

70. McCoy, *Question of Torture*, 58.

71. Author's interview with Hellyette Bess, February 7, 2011.

72. Also known as the Israeli Security Agency (ISA), Shabak, or the Shin Bet.

73. It should be noted that in recent reports we find the spellings *shabah*, *shabak*, or *shabach*, often referring specifically to the position of the detainee, handcuffed to a very low chair so that he is leaning forward.

74. Yuval Ginbar, "Routine Torture: Interrogation Methods of the General Security Service," B'Tselem, February 1998, 11.

75. Testimony of state defense attorneys before the Committee Against Torture, quoted in Ginbar, "Routine Torture," 24.

76. High Court of Justice 4025/96, "Bal'al Hanihan et al. v. General Security Service, Answer," July 2, 1996, quoted in Ginbar, "Routine Torture," 13.

77. High Court of Justice 2210/96, "Ziyad Mustafa a-Zaghel v. General Security Service," March 27, 1996, quoted in Ginbar, "Routine Torture," 13.

78. "Question of the Human Rights of All Persons Subjected to Any Form of Detention or Imprisonment, in Particular: Torture and Other Cruel, Inhuman or Degrading Treatment or Punishment," Report of the Special Rapporteur, Mr. Nigel S. Rodley, submitted pursuant to Commission on Human Rights resolution 1995/37, UN Commission on Human Rights, January 1996, para. 121.

79. HaMoked, "Torture in Secret Facility 1391: HCJ 11447/04—HaMoked: Center for the Defence of the Individual v. State of Israel (judgment rendered June 14, 2006)," February 20, 2010.

80. Yossi Wolfson, "Kept in the Dark: Treatment of Palestinian Detainees in the Petah-Tikva Interrogation Facility of the Israel Security Agency," B'Tselem/HaMoked, October 2010, 45. Observations based on the 645 charges brought by B'Tselem between 2001 and 2009.

81. Hervey, *Un monde tortionnaire*, 171.

82. Testimony of Bahjat Yamen in Noam Hoffstater, "Ticking Bombs: Testimonies of Torture Victims in Israel," Public Committee Against Torture in Israel, May 2007, 14.

83. Wolfson, "Kept in the Dark," 17.

84. Ibid., 49–50.

85. A spiritual movement that since 1999 has been considered a "heretical sect" by the Chinese government.

86. Hervey, *Un monde tortionnaire*, 125.

87. Ibid., 122–23.

88. The *laogai* (reform through work) and the *laojiao* (reeducation through work) are detention centers and work camps.

89. Hervey, *Un monde tortionnaire*. These elements are corroborated by a UN report: Manfred Nowak, "Civil and Political Rights, Including the Question of Torture and Detention. Mission to China," Economic and Social Council, Human Rights Commission, March 10, 2006.

90. Hervey, *Un monde tortionnaire*, 128.

91. McCoy, *Question of Torture*, 108.

92. Also known as bathtub torture and practiced by the French military during the Algerian War of Independence.

93. *Field Manual 34-52* describes interrogation techniques authorized by the U.S. Army, in keeping with international and domestic laws. In force since 1992, the manual was revised in 2006 following the abuses committed at the Abu Ghraib prison.

94. Ibid., 113, 121–23, 127, 134, 135, 221; Jane Mayer, "The Black Sites," *New Yorker*, August 13, 2007; Jane Mayer, "Outsourcing Torture," *New Yorker*, February 14, 2005; Department of Defense, "Memorandums: Counter-Resistance Techniques. Legal Review of Aggressive Interrogation Techniques. Legal Brief on Proposed Counter-Resistance Strategies. Request for Approval of Counter-Resistance Strategies," October 11, 2002, October 25, 2002, and November 27, 2002, www.washingtonpost.com/wp-srv/nation/documents/dodmemos.pdf (accessed March 1, 2011); Department of the Army, "Memorandum for Commander. Interrogation and Counter-Resistance Policy," September 14, 2003, www.aclu.org (accessed March 1, 2011).

95. Suzanne G. Cusick, "Music as Torture/Music as Weapon," *Transcultural Music Review*, 2006, www.sibetrans.com (accessed February 28, 2011).

96. Jane Mayer, "The Experiment," *New Yorker*, July 11, 2005; McCoy, *Question of Torture*, 126–27.

97. Mayer, "Experiment."

98. Ibid.; Mayer, "Black Sites"; Andy Worthington, "A History of Music Torture in the War on Terror," *CounterPunch*, December 15, 2008; Suzanne G. Cusick, "'You Are in a Place That Is Out of the World': Music in the Detention Camps of the 'Global War on Terror,'" *Journal of the Society for American Music* 2, no. 1 (2008): 9–10.

99. Cusick, "'You Are in a Place.'"

100. Mayer, "Black Sites."

101. Worthington, "History of Music Torture"; Human Rights Watch, "U.S. Operated Secret Dark Prison in Kabul," December 19, 2005.

102. Reprieve, "Human Cargo: Binyam Mohamed and the Rendition Frequent Flyer Programme," June 10, 2008, 36.

103. Cusick, "Music as Torture"; Jonathan Pieslak, *Sound Targets: American Soldiers and Music in the Iraq War* (Bloomington: Indiana

University Press, 2009), 86, 88, 89; Worthington, "History of Music Torture."

104. Cusick, "Music as Torture."

105. Pieslak, *Sound Targets*, 89.

106. Ibid., 169.

107. Sheila Whiteley, "Progressive Rock and Psychedelic Coding in the Work of Jimi Hendrix," in *Reading Pop: Approaches to Textual Analysis in Popular Music*, ed. Richard Middleton (Oxford: Oxford University Press, 2000), quoted in Pieslak, *Sound Targets*.

108. Cusick, "Music as Torture."

109. Pieslak, *Sound Targets*, 86.

110. Worthington, "History of Music Torture."

111. Ibid.

112. Zero dB includes musicians such as Peter Gabriel, Massive Attack, Dizzee Rascal, R.E.M., Graham Coxon, Doves, and Pearl Jam; see www.zerodb.org (accessed February 1, 2011).

113. Reprieve, "Musicians' Letter to Obama," July 16, 2009, www.reprieve.org.uk/press/2009_07_16musicionslettertoobama.

114. Department of the Army, "Army Regulation 15-6: Final Report. Investigation into FBI Allegations of Detainee Abuse at Guantanamo Bay, Cuba Detention Facility," April 1, 2005, amended June 9, 2005, 9.

115. Department of the Army, "Human Intelligence Collector Operations. FM 2-22.3 (FM 34-52)," 2006, sec. 8-21, 8-22, 8-49, 8-50, 8-51.

116. Gretchen Borchelt, Jonathan Fine, and Christian Pross, "Break Them Down: Systematic Use of Psychological Torture by U.S. Force," Physicians for Human Rights, 2005.

117. Nathaniel Raymond and Scott Allen, ed., "Experiments in Torture, Evidence of Human Subject Research and Experimentation in the 'Enhanced' Interrogation Program," Physicians for Human Rights, June 2010.

118. Jonathan H. Marks and Gregg Bloche, "The Ethics of Interrogation: The U.S. Military's Ongoing Use of Psychiatrists," *New England Journal of Medicine* 359, no. 11 (September 11, 2008): 1090.

119. McCoy, *Question of Torture*, 183.

## 5. "Hell's Bells": Medium-Frequency Sounds

1. "Hell's Bells" is a song by the Australian hard rock group AC/DC used by the U.S. Army during the Iraq War of 2003.

2. Joshua 6:20.

3. Roman Vinokur, "Acoustic Noise as a Non-Lethal Weapon," *Sound and Vibration*, October 2004, 20, based on information provided to the author by David Lubman, an acoustic consultant.

4. Wu Juncang and Zhang Qiancheng, "The Doctrine of Psychological Operations in Ancient China," *Junshi Kexue, China Military Science*, no. 5 (2002): 88–94, cited in Timothy L. Thomas, "New Developments in Chinese Strategic Psychological Warfare," *Special Warfare*, April 2003, 4.

5. Pierre Liénard, *Petite histoire de l'acoustique: Bruits, sons et musique* (Paris: Lavoisier, 2002), 38, citing Jean-François Borsarello and Agnès Robert, *Le Psychisme et la musicothérapie des Chinois* (Paris: Éditions de la Maysnie, 1983), 95.

6. Liénard, *Petite histoire*, 376.

7. Tonnfort company website, www.tonnfort.com/effaroucheur.html (accessed May 22, 2011).

8. Vinokur, "Acoustic Noise," 19–20, according to Dimitri Obolensky, ed., *The Heritage of Russian Verse* (Bloomington: Indiana University Press, 1976).

9. Olive Lewin, *Rock It Come Over: The Folk Music of Jamaica; with Special Reference to Kumina and the Work of Mrs. Imogene "Queenie" Kennedy* (Kingston: University of the West Indies Press, 2001), 158, quoted in Steve Goodman, *Sonic Warfare: Sound, Affect and the Ecology of Fear* (Cambridge, MA: MIT Press, 2010), 65.

10. Goodman, *Sonic Warfare*, 65–66.

11. Murray Schafer, *The Soundscape: Our Sonic Environment and the Tuning of the World* (Rochester, NY: Destiny Books, 1977, 1994), 178.

12. In 2010, psyops were renamed MISO (military support to information operations), allegedly to undo a reputation for being "underhanded and unethical," according to Colonel Curtis Boyd in "The Future of MISO," *Special Warfare*, January–February 2011.

13. The two other branches are operational psyops, having an impact on the evolution of the war, and strategic psyops, which is international in scope and intervenes via newspapers, radio, the Internet, or television (Department of the Army, Department of the Navy, United States Marine Corps, Department of the Air Force, United States Coast Guard, "Joint Publication 3-53. Doctrine for Joint Psychological Operations," September 5, 2003, I-4).

14. Vinokur, "Acoustic Noise," 19.

15. Marie-Catherine Villatoux and Paul Villatoux, "Voices from Heaven," www.psywar.org, May 25, 2008.

16. Mark Lloyd, *The Art of Military Deception* (London: Leo Cooper, 1997), 146, quoted in Jonathan Pieslak, *Sound Targets: American Soldiers and Music in the Iraq War* (Bloomington: Indiana University Press, 2009), 80.

17. Christoph Cox, "Edison's Warriors," *Cabinet*, no. 13 (spring 2004); Goodman, *Sonic Warfare*, 41.

18. Cox, "Edison's Warriors."

19. So named by its founder, Major General George O. Squier, by combining "music" and "Kodak."

20. The film came out in 1940, on the occasion of which the stereophonic system developed by Disney was named Fantasound.

21. James Tobias, "Composing for the Media: Hanns Eisler and Rockefeller Foundation Projects in Film Music, Radio Listening, and Theatrical Sound Design," Rockefeller Archives, 2009, 8, 66, 69. This study portrays

Burris-Meyer as being generally interested in emotional manipulation and crowd control by means of audio (8).

22. Ibid.

23. Philip Gerard, *Secret Soldiers: How a Troupe of American Artists, Designers, and Sonic Wizards Won World War II's Battles of Deception Against the Germans* (New York: Penguin, 2002), 105, quoted in Goodman, *Sonic Warfare*, 42.

24. Gerard, *Secret Soldiers*, 111, quoted in Goodman, *Sonic Warfare*, 214.

25. Cox, "Edison's Warriors."

26. Rick Beyer, "Now Hear This," excerpt from *The Ghost Army*, www .ghostarmy.org (accessed March 15, 2011); Gerard, *Secret Soldiers*, 115, quoted in Goodman, *Sonic Warfare*, 214.

27. Cox, "Edison's Warriors."

28. Ibid. The army denied the existence of this operation on several occasions. The facts are now documented by several sources, in addition to Gerard, *Secret Soldiers*, and the documentary by Rick Beyer, *The Ghost Army*; Cox mentions as references Jonathan Gawne, *Ghosts of the ETO: American Tactical Deception Units in the European Theater, 1944–1945* (Havertown, MD: Casemate, 2002), and Jack Kneece, *Ghost Army of World War II* (Gretna, LA: Pelican, 2001).

29. Gerard, *Secret Soldiers*, 105–6, quoted in Goodman, *Sonic Warfare*, 43.

30. Tobias, "Composing for the Media," 8, 73.

31. Cox, "Edison's Warriors."

32. Villatoux and Villatoux, "Voices from Heaven."

33. Pierre Pahlavi, "La guerre politique: Une arme à double tranchant. Montée en puissance et déclin de la contre-insurrection française en Algérie," *Revue militaire canadienne*, winter 2007–8.

34. Francis Ford Coppola, dir., *Apocalypse Now*, 1979.

35. Rex Applegate, *Riot Control: Material and Techniques* (Harrisburg, PA: Stackpole, 1969), 270–71.

36. Villatoux and Villatoux, "Voices from Heaven."

37. Named for Wandering Souls Day (VuLan), an annual Buddhist ceremony during which homage is paid to the dead. A sound excerpt of the cassette was placed online in an article by Sergeant Major Herbert A. Friedman, "The Wandering Soul: PSYOP Tape of Vietnam," n.d., www .pcf45.com/sealords/cuadai/wanderingsoul.html (accessed March 13, 2011).

38. Lieutenant Colonel Rieka Stroh and Major Jason Wendell, "A Primer for Deception Analysis: Psychological Operations. Target Audience Analysis," *IO Sphere*, fall 2007, 46.

39. Members of PSYOP 6th battalion, quoted in Friedman, "Wandering Soul."

40. Duane Yeager, "Winning Vietnamese Minds Was What the US Army's 4th Psychological Operations Group Was All About," *Vietnam Magazine*, December 1990, quoted in Friedman, "Wandering Soul."

41. Bill Rutledge, quoted in Friedman, "Wandering Soul."

42. An officer in the 10th battalion of PSYOP, quoted in Friedman, "Wandering Soul."

43. Villatoux and Villatoux, "Voices from Heaven."

44. Tommy Franks, *American Soldier* (New York: HarperCollins, 2004), 157, quoted in Christopher J. Lamb, "Review of Psychological Operations Lessons Learned from Recent Operational Experience," National Defense University Press, Washington, DC, September 2005, 82.

45. Major Ed Rouse, "Gulf War Loudspeaker Victories," n.d., www .psywarrior.com/loudspeaker.html (accessed March 16, 2011).

46. Ronald H. Cole, "Operation Just Cause: The Planning and Execution of Joint Operations in Panama, February 1989–January 1990," Joint History Office, 1995, 59.

47. Pieslak, *Sound Targets*, 82.

48. Lane Degregory, "Iraq 'n' Roll," *St. Petersburg Times*, November 21, 2004.

49. Cole, "Operation Just Cause," 59.

50. Ibid., 60.

51. Degregory, "Iraq 'n' Roll."

52. David A. Koplow, "Tangled Up in Khaki and Blue: Lethal and Non-Lethal Weapons in Recent Confrontations," *Georgetown Journal of International Law* 36 (2005): 761.

53. Richard Scruggs, Victor Gonzalez, Steven Zipperstein, et al., "Report to the Deputy Attorney General on the Events at Waco, Texas, February 28 to April 19, 1993," Department of Justice, October 8, 1993, sec. II.

54. Neil Davison, "Occasional Report #2. The Development of 'Non Lethal' Weapons During the 1990's," BNLWRP, March 2007, 4–5; Psychotechnology Research Institute of Dr. Smirnov website, www.psycor.info (accessed December 23, 2010)—the site is now exclusively in Russian, but an old version of it is available in English at web.archive.org/web/20070706021955 /http://int.psycor.ru/Mayn.php?prka (accessed May 22, 2011).

55. Scruggs et al., "Report to the Deputy Attorney General," sec. III-B.

56. Alan A. Stone, "Report and Recommendations Concerning the Handling of Incidents Such as the Branch Davidian Standoff in Waco Texas," November 10, 1993, sec. IV-B.

57. Testimony of C.J. Grisham, quoted in Pieslak, *Sound Targets*, 51.

58. Pieslak, *Sound Targets*, 49, 50, 55.

59. Ibid., 83.

60. Bing West, *No True Glory: A Frontline Account of the Battle of Fallujah* (New York: Bantam, 2005), 176, quoted in Pieslak, *Sound Targets*, 84–85.

61. Degregory, "Iraq 'n' Roll."

62. Cusick, "Music as Torture."

63. Degregory, "Iraq 'n' Roll."

64. Jonathan B. Keiser and Mark C. Engen, "Adapting the Vehicle Mounted Tactical Loudspeaker System to Today's Operational Environment," thesis, Naval Postgraduate School, Monterey, CA, December 2006, 34.

65. Sergeant Major Herbert A. Friedman, "The 361st Psychological Operations Company in Iraq," May 18, 2005, www.psywarrior.com (accessed March 2, 2011).

66. Power Sonix website, www.powersonix.com (accessed January 30, 2011).

67. That is to say, modular amplifiers and loudspeakers, classified as a function of their use in the field: whether carried on backs, by vehicles, by airplanes, or on ships.

68. U.S. Special Operations Command (USSOCOM), "Sources Sought Notice: PSYOP. Family of Loud-speakers II (FOL2). Reference Number H92222-05-FOL2-LOUDSPEAKERS," *Federal Business Opportunities Daily*, June 8, 2005.

69. Department of Defense, "RDT&E Budget Item Justification. Project no. PE 1160488BB SOF PSYOP/D476," February 2008, 4.

70. A human rights activist expressing his opposition to the Israeli Shofar, quoted in Mitch Potter, "Israelis Unleash Scream at Protest," *Toronto Star*, June 6, 2005.

71. Jürgen Altmann, "Acoustic Weapons, a Prospective Assessment: Sources, Propagation and Effects of Strong Sound," Occasional Paper no. 22, Cornell University Peace Study Programs, May 1999, 55.

72. Ibid., 23, 27, 55.

73. Applegate, *Riot Control*, 271.

74. Carol Ackroyd, Karen Margolis, Jonathan Rosenhead, and Tim Shallice, *The Technology of Political Control* (Harmondsworth, UK: Penguin, 1977), 224. The Curdler cost around £2,000 at the time.

75. Ibid., 224.

76. Reports of Harry Diamond Laboratories, cited in Applied Research Laboratories, "Non-Lethal Swimmer Neutralization Study," University of Texas, Austin, G2 Software Systems Inc., May 2002, 55.

77. ARL, "A History of the Army Research Laboratory," August 2003, 3.

78. Harry Diamond Laboratories, "Disperse: A Survey of Relevant Literature and Research Activities," U.S. Army Materiel Command, 1975, quoted in Applied Research Laboratories, "Non-Lethal Swimmer Neutralization Study," 17.

79. Ibid., 29.

80. Applied Research Laboratories, "Non-Lethal Swimmer Neutralization Study," 17.

81. Picatinny Arsenal, "ARDEC Exploring Less-than-Lethal Munitions," U.S. Army news release, quoted in William Arkin, "Acoustic Anti-Personnel Weapons: An Inhuman Future?" *Medicine, Conflict and Survival* 14 (1997): 321.

82. SARA, "Selective Area/Facility Denial Using High Power Acoustic Beam Technology," Phase 1 SBIR Final Report, March 10, 1995 (revised February 13, 1996), quoted in Arkin, "Acoustic Anti-Personnel Weapons," 318.

83. Altmann, "Acoustic Weapons," 28.

84. "High Power Acoustic Technology," presentation on the SARA website, www.sara.com/DE/high_power_acoustics/high_power_acoustic_tech.html (accessed March 19, 2011).

85. Charles T. Freund, "Aversive Audible Acoustic Devices," U.S. Army TACOM-ARDEC, Advanced Armaments Technologies, Close Combat

Armaments Center, NDIA 2000 Joint Services Small Arms Symposium, August 29, 2000.

86. Harry Moore, "Multi-Sensory Deprivation Land-Mine," U.S. Army TACOM-ARDEC, Advanced Armaments Technology Team, NDIA Small Arms Symposium, May 14, 2002.

87. Beck, Short, Van Meenen, and Servatius, "Suppression Through acoustics," *Proceedings of SPIE* 6219 (2006), quoted in Neil Davison, "*Non-Lethal*" *Weapons* (Houndmills, UK: Palgrave Macmillan, 2009), 200.

88. China Analyst, "Top 10 Best-Performing Defense Stocks in 2010," January 3, 2011.

89. Piezoelectricity, electricity resulting from pressure, is the charge that accumulates in certain solid materials in response to applied mechanical stress. "The piezoelectric effect is understood as the linear electromechanical interaction between the mechanical and the electrical state in crystalline materials with no inversion symmetry. The piezoelectric effect is a reversible process in that materials exhibiting the direct piezoelectric effect (the internal generation of electrical charge resulting from an applied mechanical force) also exhibit the reverse piezoelectric effect (the internal generation of a mechanical strain resulting from an applied electrical field)" ("Piezoelectricity," *Wikipedia*, en.wikipedia.org/wiki/Piezo electricity). Piezoelectric materials are used, notably, in the acoustic domain to convert sound waves into electricity or to transform an electric signal into an acoustic signal (in the case of loudspeakers).

90. Ian Sample, "Pentagon Considers Ear-Blasting Anti-Hijack Gun," *New Scientist*, November 14, 2001.

91. Ibid.

92. Henry S. Kenyon, "Noisemakers Called to Arms," *Signal Magazine*, July 2002.

93. James R. Jauchem and Michael C. Cook, "High-Intensity Acoustics for Military Non-Lethal Applications," *Military Medicine*, February 2007.

94. Technical specifications of different versions of the LRAD on the LRAD Corp website, www.lradx.com (accessed March 18, 2011); Jürgen Altmann, "Millimetre Waves, Lasers, Acoustics for Non-Lethal Weapons? Physics Analyses and Interferences," DSF-Forschung, 2008, 45.

95. William Arkin, "The Pentagon's Secret Scream: Sonic Devices That Can Inflict Pain—or Even Permanent Deafness—Are Being Deployed," *Los Angeles Times*, March 7, 2004.

96. Altmann, "Millimetre Waves," 52.

97. Neil Davison and Nick Lewer, "Research Report No. 8," BNLWRP, March 2006, 33; Sharon Weinberger, "Acoustic Weapon Hits Georgian Protesters," *Wired*, November 14, 2007; Andrew Darby, "Whalers Attack Activists at Sea," *The Age*, February 6, 2009; Raffi Khatchadourian, "Street Fight on the High Seas," *New Yorker*, January 12, 2010; Clean Clothes Campaign, "Leaders of Peaceful Protest Against Triumph Threatened with Arrest," September 7, 2009, www.cleanclothes.org (accessed March 19, 2011); "Testés en Irak, appliqués au G20: Les nouveaux cannons anti-émeute,"

Numéro Lambda, October 6, 2009, numerolambda.wordpress.com (accessed March 19, 2011); David Hambling, "Sonic Warfare Erupts in Pittsburgh, Honduras," *Wired*, September 25, 2009; "DHS Helps Local Police Buy Military-Style Sonic Devices," *Washington Times*, October 1, 2009.

98. Chuck McCutcheon, "Military's Needs Speed Development of New Non-Lethal Weapons," Newhouse News Service, June 9, 2004.

99. Friedman, "361st Psychological Operations Company."

100. The manufacturer and the military press have produced various information sheets on the question. See, for example, the "free tribune" by Robert Putnam, who handled public relations at LRAD Corp, "LRAD No Weapon," *Pittsburgh Tribune-Review*, October 27, 2009.

101. David Hambling, "US 'Sonic Blasters' Sold to China," *Wired*, May 15, 2008.

102. Altmann, "Millimetre Waves," 44, 52.

103. In 2004, the New York police paid around $25,000, and in 2009, the San Diego police paid about $22,000. Neil Davison and Nick Lewer, "Research Report No. 6," BNLWRP, October 2004, 11; *Washington Times*, "DHS Helps Local Police."

104. *CCLA v. Toronto Police Service*, 2010 ONSC 3525, June 6, 2010, 25, 26, 31, 33, 37, 43; "Quatre cannons soniques à Toronto," Radio-Canada, May 21, 2010; Mike Smith, "Up to Our Ears in New Police Tech," Toronto OpenFile, February 4, 2011, toronto.openfile.ca (accessed February 10, 2011); Catherine Porter, "A Loud and Clear Lesson in Police Power," *Toronto Star*, February 4, 2011; Canadian Civil Liberties Association, "CCLA Concerned About Use of Sound Cannons Prior to Their Review by Ontario Gov," news release, February 6, 2011.

105. Defense Contracting Command Washington, "FBO Daily Issue of February 22, 2004. FBO #0818," *Federal Business Opportunities*, February 20, 2004.

106. David Hambling, "Say Hello to the Goodbye Weapon," *Wired*, May 12, 2006; "Perfect Weapon Is Perfectly Useless," *Strategy Page*, July 30, 2010; Davison and Lewer, "Research Report No. 6," 10–11.

107. Joint Non Lethal Weapons Directorate, "Distributed Sound and Light Array (DSLA)," December 2010, www.jnlwp.usmc.mil (accessed April 18, 2011).

108. Sandra Jontz, "Troops in Iraq to Get Combined Lethal/Non-Lethal Weapons System," *Stars and Stripes*, September 14, 2004.

109. Government Accountability Office, "Matter of IMLCORP LLC; Wattre Corporation. File B-310582; B-310582.2; B-310582.3; B-310582.4; B-310582.5," January 9, 2008; Wattre Corporation, "World Record Sound Information," March 2007, Wattre Corporation website, www.ultra-hyperspike.com (accessed December 20, 2010).

110. See the manufacturer's website, www.getmad.com (accessed February 21, 2011).

111. Lee Bzorgi, quoted by Frank Munger, "Creator: Banshee II Non-Lethal Weapon Assaults Only Ears," Knoxnews, September 1, 2009, blogs.knoxnews.com/munger/ (accessed December 20, 2010).

112. Frank Munger, "What Does Banshee II Look Like?" Atomic City Underground, Knoxnews, September 10, 2009, www.knoxnews.com (accessed December 20, 2010).

113. See the presentation of Inferno Intenso on the website www.inferno.se (accessed March 20, 2011). Indusec's "sound barriers" can be used in conjunction with its "light barriers," namely, stroboscopes that give off intermittent flashes of light.

114. Applied Research Laboratories, "Non-Lethal Swimmer Neutralization Study," 30–31.

115. Davison, *"Non-Lethal" Weapons*, 202.

116. American Technology Corporation, "Annual Report (10-K)," July 1, 2008.

117. Jeremy Page, "Available: Chinese Tech for Putting Down Protests," *Wall Street Journal*, February 26, 2011; CETC, "Introduction to Directed High-Intensity Acoustic Low-Lethal Weapons for Police," www.cetci.com.cn (accessed February 26, 2011).

118. Mitch Potter, "Israelis Unleash Scream at Protest," *Toronto Star*, June 6, 2005.

119. Ibid.

120. Technical specifications of the Shofar from the manufacturer's website, www.eord.co.il/category.asp?id=171 (accessed January 29, 2011).

121. Bex Tyrer, "Bil'in: Une leçon de résistance créative," Association France-Palestine Solidarité, October 7, 2005, www.france-palestine.org (accessed December 23, 2010).

122. Potter, "Israelis Unleash Scream."

123. Hacène Belmessous, *Opération banlieues: Comment l'État prépare la guerre urbaine dans les cités françaises* (Paris: La Découverte, 2010), 66–67.

124. Benjamin Ferran, "Le 'cannon à son,' nouvelle arme contre les manifestants," *Le Figaro*, September 30, 2009.

125. All quotes and technical information come from the description of the Mosquito MK4 and of the Mosquito mood calming system on the manufacturer's website, www.compoundsecurity.co.uk (accessed February 21, 2011).

126. Erik Lacitis, "Hearing Bing Sing for Teens, a Real Earful," *Seattle Times*, July 9, 1999; "Manilow to Drive Out 'Hooligans,'" BBC News, June 5, 2006; AM Archive, "Pink Lighting and Bing Crosby," ABC, July 8, 1999.

127. Associated Press, "Annoying 'Mosquito' Noise Keeps Students Moving," March 15, 2009.

128. Children's Commissioner, "Buzz Off Campaign," 2008, www.childrenscommissioner.gov.uk (accessed March 21, 2011); "Sonic Youth Weapon Should Be Banned," Indymedia Irlande, February 12, 2008, www.indymedia.ie (accessed March 21, 2011).

129. Government of Ireland, Non-Fatal Offences Against the Person Act, 1997, article 2.

130. Article R 1334-31 of the public health code.

131. IPB's French website in 2009, web.archive.org/web/20090426032803/http://www.ibpfrance.fr/ultrasonbeethoven/ (accessed

March 21, 2011); Emeline Cazi, Pascale Egré, and Claudine Proust, "Le répulsif antijeunes qui fait débat," *Le Parisien*, April 24, 2008; "Le boîtier anti-jeunes devant le juge," *Le Parisien*, April 24, 2008; Family Court of Saint-Brieuc, "Association Val Tonic, Brigitte G., Franck H. v. Philippe L. No RG: 08/00106. Ordonnance de référé," April 30, 2008.

132. Belga, "Motion anti-Mosquito à Bruxelles," *La Libre Belgique*, April 26, 2008; Belga, "Pas d'interdiction du Mosquito," *La Libre Belgique*, April 2, 2008; National Physical Laboratory, "Test Report: 'Mosquito' Sound Source," December 7, 2005; Royal College of Pediatrics and Child Health, "Mosquito High Frequency Sound Deterrent and Teenagers/Children," Advocacy Report, November 2007; Home Department, "Antisocial Behaviour: Mosquito Device," May 24, 2007.

133. Jacques Chatillon, "Limites d'exposition aux infrasounds et aux ultrasounds. Étude bibliographique," INRS, Hygiène et sécurité du travail, 2Q 2006, 75.

134. An executive at ATC/LRAD Corp speaking of the HSS, in John Gartner, "Point 'n' Shoot Sound Makes Waves," *Wired*, February 21, 2002.

135. Elwood Norris, quoted in Marshall Sella, "The Sound of Things to Come," *New York Times Magazine*, March 23, 2003.

136. Information published on the dedicated site www.audiospotlights .com (accessed June 14, 2010; taken down March 22, 2011), the primary goal of which was to show, through numerous comparison tables, the advantages of the Audio Spotlight compared to the HSS.

137. Andrew Hampp, "Hear Voices? It May Be an Ad," *Advertising Age*, December 10, 2007.

138. Directional Sound, "Use as a Repellent," www.directional-sound.eu /en/application-of-directional-sound/use-as-repellent (accessed March 21, 2011).

139. Sella, "Sound of Things."

140. Gartner, "Point 'n' Shoot Sound."

141. DARPA, "RDT&E Budget Item Justification. PE 0602702E Tactical Technology," 2010, 36; Sharon Weinberger, "A Voice Only You Can Hear: DARPA's Sonic Projector," *Wired*, June 5, 2007.

142. INRS, "Guide pour l'établissement de limites d'exposition aux champs électriques, magnétiques et électromagnétiques," Cahiers de notes documentaires, "Hygiène et sécurité du travail," no. 182, 1Q 2001, 32.

143. Department of the Army, "Bioeffects of Selected Non-Lethal Weapons," 2006, 6.

144. Davison, *"Non-Lethal" Weapons*, 175.

## 6. "No Matter What Your Purpose Is, You Must Leave": The Sound of Power

1. Declaration broadcast in a loop, via an LRAD, by the police of Pittsburgh to the G20 protesters in August 2009, alternating with the device's warning signal.

2. Chronicler Catherine Porter on the use by the Canadian police of the LRAD, allegedly as a "tool of communication": "If you need a high-powered megaphone, get a high-powered megaphone. Don't get one attached to an auditory machine-gun."

3. Quoted in Steve Wright, "The New Technologies of Political Repression: A New Case for Arms Control?" *Philosophy and Social Action*, July–December 1991.

4. Michel Foucault, *Histoire de la sexualité, vol. 1: La Volonté de savoir* (Paris: Gallimard, 1976), 183.

5. Georges-Henri Bricet des Vallons, "L'arme non létale dans la stratégie militaire des États-Unis: Imaginaire stratégique et genèse de l'armement," *Cultures et Conflits*, no. 67 (fall 2007), conflits.revues.org (accessed March 13, 2011).

6. Sun Tzu, *The Art of War*, chap. 3.

7. Letter from Major General W.C. Wyman to Major General Lauris Norsted, dated July 22, 1947, quoted by Colonel Alfred H. Paddock, "Psychological and Unconventional Warfare, 1941–1952: Origins of a 'Special Warfare' Capability for the United States Army," U.S. Army War College, November 1979, and referenced in Colonel Paul E. Valley and Major Michael A. Aquino, "From PSYOP to Mindwar: The Psychology of Victory," Headquarters, 7th Psychological Operations Group, U.S. Army Reserve, 1980, updated Nov. 2003, 5.

8. Neil Davison, "The Contemporary Development of 'Non-Lethal' Weapons," BNLWRP, May 2007, 24.

9. Lab presentation on its website, njms2.umdnj.edu/smbiweb/index .html (accessed April 22, 2011).

10. Neil Davison, "The Early History of 'Non-Lethal' Weapons," BNLWRP, December 2006, 12.

11. Presentation of civilian application of the LRAD on the website www.lradx.com/site/content/view/296/110 (accessed May 5, 2011).

12. Steve Wright, "Violent Peacekeeping: The Rise and Rise of Repressive Techniques and Technologies," Praxis Centre, Leeds Metropolitan University, January 28, 2005.

13. Bricet des Vallons, "L'arme non létale."

14. Ibid.

15. Neil Davison, *"Non-Lethal" Weapons* (Houndmills, UK: Palgrave Macmillan, 2009), 46, 58; Steve Wright, "Merchants of Repression," *GSC Quarterly*, no. 12 (spring 2004).

16. Bricet des Vallons, "L'arme non létale." All citations in this paragraph come from this article.

17. Ibid.

18. ONU, Convention on Prohibitions or Restrictions on the Use of Certain Conventional Weapons Which May Be Deemed to Be Excessively Injurious or to Have Indiscriminate Effects, Geneva, October 10, 1980.

19. David A. Koplow, "Tangled Up in Khaki and Blue: Lethal and Non-Lethal Weapons in Recent Confrontations," *Georgetown Journal of International Law* 36 (2005): 796–99.

20. Davison, "Early History"; Wright, "New Technologies"; Steve Wright, "Civilising the Torture Trade," *The Guardian*, March 13, 2003.

21. Timothy L. Thomas, "The Mind Has No Firewall," *Parameters*, spring 1998, 84–92.

22. Cited in Wright, "New Technologies."

23. On this note, one may listen to the sound documentary *Mosquito*, by Olivier Toulemonde, in which one hears a Kafkaesque exchange between people bothered by the frequencies of the device illegally installed by a bank in Brussels and the police who don't hear the sounds and remain incredulous: www.olivier-toulemonde.com/mosquito_053.htm (accessed March 21, 2011).

24. Ibid.

25. Jürgen Altmann, "Acoustic Weapons, a Prospective Assessment: Sources, Propagation and Effects of Strong Sound," Occasional Paper no. 22, Cornell University Peace Study Programs, May 1999, 58–59.

26. Robin Coupland, "The Sirus Project," Comité international de la Croix-Rouge, 1997, 24

27. Ibid., 23.

28. Brian Rappert, "A Framework for the Assessment of Non-Lethal Weapons," *Medicine, Conflict and Survival* 20 (2004): 41, 42, 46.

29. Wright, "Violent Peacekeeping."

30. A video game.

31. Major Mark R. Thomas, "Non-Lethal Weaponry: A Framework for Future Integration," Air Command and Staff College, Air University, Maxwell Air Force Base, AL, April 1998, 12–13.

32. Bricet des Vallons, "L'arme non létale."

33. Suzanne G. Cusick, "Music as Torture/Music as Weapon," *Transcultural Music Review*, 2006, www.sibetrans.com (accessed February 28, 2011).

34. Quoted in Lane Degregory, "Iraq 'n' Roll," *St. Petersburg Times*, November 21, 2004.

35. Quoted in Andy Worthington, "A History of Music Torture in the War on Terror," *CounterPunch*, December 15, 2008.

36. Cusick, "Music as Torture."

37. Françoise Sironi, "Les mécanismes de destruction de l'autre," in *L'Empathie*, ed. Alain Berthoz and Gérard Jorland (Paris: Odile Jacob, 2004), 228, quoted in Michel Terestchenko, *Du bon usage de la torture, ou comment les démocraties justifient l'injustifiable* (Paris: La Découverte, 2008), 159.

38. Wright, "Violent Peacekeeping."

39. Definition by Colonel Alexander in a book co-written with Major Richard Groller and Janet Morris (and with a preface by spy novelist Tom Clancy), *The Warrior's Edge: Front-line Strategies for Victory on the Corporate Battlefield* (New York: William Morrow, 1990), quoted in Steven Aftergood, "The 'Soft Kill' Fallacy," *Bulletin of the Atomic Scientists* 50, no. 5 (September–October 1994): 40–45.

40. Bricet des Vallons, "L'arme non létale."

41. Lieutenant Colonel John B. Alexander, "The New Mental Battlefield: 'Beam Me Up, Spock,'" *Military Review* 60, no. 12 (December 1980): 52.

42. Altmann, "Acoustic Weapons," 2.

43. Nick Lewer and Steven Schofield, *Non-Lethal Weapons: A Fatal Attraction?* (London: Zed Books, 1997), 36, cited in Neil Davison, "The Development of 'Non-Lethal' Weapons During the 1990's," BNLWRP, March 2007, 13.

44. Wright, "Violent Peacekeeping."

45. Description of the app Sound Grenade for iPod on the iTunes website, itunes.apple.com/us/app/sound-grenade/id301687625?mt=8 (accessed March 13, 2011).

46. TeenBuzz website, www.teenbuzz.org (accessed April 25, 2011).

47. Friedrich Kittler, *Gramophone, Film, Typewriter*, trans. Geoffrey Winthrop-Young and Michael Wutz (Stanford, CA: Stanford University Press, 1999), 97, quoted in Steve Goodman, *Sonic Warfare: Sound, Affect and the Ecology of Fear* (Cambridge, MA: MIT Press, 2010), 32.

48. James Tobias, "Composing for the Media: Hanns Eisler and Rockefeller Foundation Projects in Film Music, Radio Listening, and Theatrical Sound Design," Rockefeller Archives, 2009, 68.

49. Geoffrey Winthrop-Young, "Drill and Distraction in the Yellow Submarine: On the Dominance of War in Friedrich Kittler's Media Theory," *Critical Inquiry* 28, no. 4 (summer 2002): 830.

50. Kittler, *Gramophone, Film, Typewriter*, 97, quoted in Goodman, *Sonic Warfare*, 209–10.

51. David Hambling, *Weapons Grade: How Modern Warfare Gave Birth to Our High-Tech World* (New York: Carroll & Graf, 2005), 250.

52. Timothy Lenoir, "Programming Theaters of War: Gamemakers as Soldiers," in *Bombs and Bandwidth: The Emerging Relationship between Information Technology and Security*, ed. Robert Latham (New York: The New Press, 2003).

53. Ibid.

54. Stephen Stockwell and Adam Muir, "The Military-Entertainment Complex: A New Facet of Information Warfare," *Fibre Culture*, no. 1 (2003).

55. Paula P. Henry, Bruce E. Amrein, and Mark A. Ericson, "The Environment for Auditory Research," *Acoustics Today* 5, no. 3 (July 2009): 9.

56. Associated Press, "Audio Battlefield Aims to Help Prepare New Troops," May 6, 2011.

57. Quoted in Duncan Geere, "Surround-Sound 'Audio Battlefield' Reproduces Cacophony of War," *Wired*, May 11, 2011.

58. Joseph D. Grimes, "Modeling Sound as a Non-Lethal Weapon in the COMBAT XXI Simulation Model," thesis, Naval Postgraduate School, Monterey, CA, June 2005, i.

59. Olivier Razac, "L'utilisation des armes de neutralisation momentanée en prison: enquête auprès des formateurs de l'ENAP [École nationale d'administration pénitentiaire]," thematic files from CIRAP, no. 5, July 2008, 27.

60. Jean-Baptiste Bernard, "Olivier Razac: 'Penser les nouvelles formes de violence politique basées sur la neutralization,'" Article XI, December 22, 2010, www.articlell.info (accessed December 23, 2010).

61. Goodman, *Sonic Warfare*, 64.

62. Davison, *"Non-Lethal" Weapons*, 205.

63. Noel García López, "Alarmas y sirenas: sonotopías de la conmoción cotidiana" in *Espacios sonoros, tecnopolítica y vida cotidiana* (Barcelona: Orquesta Del Caos–Institut Català d'Antropologia, 2005), 18.

64. Dr. John M. Kenny, Dr. Clark McPhail, Dr. Peter Waddington, et al., "Crowd Behavior, Crowd Control, and the Use of Non-Lethal Weapons," Human Effects Advisory Panel, INLDT, Penn State University, January 1, 2001.

65. Mike Davis, *City of Quartz: Excavating the Future in Los Angeles* (London: Verso, 1990), 241–42.

66. Benches that are inclined or separated by armrests to make it difficult to lie down or even sit down; poles, spikes on the edges of fountains, grilles, barriers of bushes, or blocks of cement with pebbles cemented in.

67. García López, "Alarmas y sirenas," 20–21.

68. Ibid., 22.

69. Ibid., 24.

70. Davis, *City of Quartz*, 269.

71. *New Scientist*, August 24, 1994, 18, quoted in Mike Davis, *The Ecology of Fear: Los Angeles and the Imagination of Disaster* (New York: Vintage, 1999), 367.

72. Davis, *Ecology of Fear*, 368.

73. Goodman, *Sonic Warfare*, 64.

74. EORD website, www.eord.co.il (accessed January 29, 2011).

75. CETC website, www.cetci.com.cn (accessed May 24, 2011).

76. European Commission, "Towards a More Secure Society and Increased Industrial Competitiveness. Security Research Projects Under the 7th Framework Program for Research," May 2009, 3.

77. Brian Martin and Steve Wright, "Looming Struggles over Technology for Border Control," *Journal of Organizational Transformation and Social Change* 3, no. 1 (2006): 95–107.

78. Wright, "Merchants of Repression."

79. Goodman, *Sonic Warfare*, xv.

80. Jean-Pierre Garnier, "L'espace public réenchanté," *Le Monde libertaire*, December 11–17, 2008.

## Conclusion: "A Passionate Sound Gesture"

1. Presentation by the collective Escoitar of its sound play *Sonic Weapons*, produced in 2009 and available at www.artesonoro.org/sonic-weapons/ (accessed April 28, 2011)

2. Murray Schafer, *The Soundscape: Our Sonic Environment and the Tuning of the World* (Rochester, NY: Destiny Books, 1977, 1994), 3.

3. William Burroughs, *Electronic Revolution* (New York: Left Bank Books, 1971).

4. Josephina Bosma, "Trembling Structures—Mark Bain," August 8, 1999, www.nettime.org/Lists-Archives/nettime-l-9908/msg00023.html (accessed May 8, 2011).

5. Steve Goodman, *Sonic Warfare: Sound, Affect and the Ecology of Fear* (Cambridge, MA: MIT Press, 2010), 76.

6. Bosma, "Trembling Structures."

7. Mark Oliver, "The Day the Earth Screamed," *The Guardian*, February 13, 2004; Molly Hankwitz and David Cox, "Interview with Mark Bain," Artists' Television Access, January 3, 2000; Bosma, "Trembling Structures."

8. Goodman, *Sonic Warfare*, 78, 79.

9. Escoitar, *Sonic Weapons*. Quotes in this paragraph are taken from the website.

# INDEX

Abel, Ben, 101, 103–4
abeng (cow horn), 93
Abu Ghraib prison, 171n93
AC/DC (band), 101, 103, 172n1
"acoustic bazooka," 109–10
"acoustic beam" weapons, 33, 35, 105
"acoustic blasters," 34, 52, 53, 108, 160n58
"acoustic bullet," 31, 32, 110
"acoustic fences." *See* sound barriers and fences
"acoustic mines," 32, 109
acoustic signatures, 6, 51, 141
Action by Christians for the Abolition of Torture, 83
Action directe (group), 67, 81
Active Denial System (ADS), 113
adolescents. *See* teenagers
advertising, 121, 122, 147
Afghanistan, 84
Ahmed, Ruhal, 88–89
"air cannons," 43, 46, 164n49
Air Force Research Laboratory (AFRL), 33, 34
airplanes, 55–57, 95, 97, 98, 99, 128. *See also* anti-aircraft weapons; hijackers
"alarm" (word), 20
alarms, 9, 147. *See also* sirens
Alexander, John, 138
Algeria, 99, 171n92
Allan Memorial Institute, 76–77
Allen International, 30
Alsetex, 58, 62–63
ALS Technologies, 60
Altmann, Jürgen, 3, 14–15, 22, 27–28, 31, 108, 139; on amplitude

limits, 133; on auditory annoyance, 106; on infrasound production, 30; on LRAD, 111
American Medical Association, 90
American Psychiatric Association, 76, 77, 90
American Psychological Association, 90
American Technology Corporation, 109
amphitheaters and concert halls, 17–18
amplification, 18, 157n19, 158n22
amplitude, 11, 13–16, 22, 43, 55; of deterrent devices, 110, 114, 116, 117, 120; of grenades, 60, 63; of helicopter harassment, 100; limits recommended, 133; of musical harassment, 104. *See also* pain threshold
anechoic chambers. *See* soundproof rooms and cells
Angliss, Sarah, 40–41, 42
animal control, 53, 109, 115, 117, 122, 128
animal experimentation, 34, 36, 52, 90, 109
animals: echolocation by, 18; hearing of, 9, 157n30; sounds of, 13, 97, 99, 157n20
anti-aircraft weapons, 45–46
anticommunism, 67–68
anti-noise headsets and helmets, 19, 110–11
anti-noise regulations, 64–65, 119
antipiracy technology, 105, 111

186

anti-swimmer technology. *See* swimmer deterrence

*Apocalypse Now* (film), 99

Applegate, Rex, 58, 126; *Riot Control*, 29, 57, 73, 99, 107

Applied Electro Mechanics, 107

Applied Research Laboratories, University of Texas, 37, 51

Applied Research Laboratory, Pennsylvania State University, 113

architectural barriers, 146, 148, 184n66

architecture, vibrational. *See* "vibrational architecture"

Arkin, William, 31

"area denial," 51, 52, 59, 105, 117–20, 145–50, 152

Army Armament Research, Development and Engineering Center (ARDEC), 32, 36, 47, 108–9, 127, 129

Army Research Laboratory (ARL), 33–34, 47, 107, 142

Armytec, 53–54

artillery, 43, 45–46

asymmetric conflict, 129–30

ATC Corporation, 121

atmospheric pressure, 43, 45

audibility, 9, 13, 23, 156n8; of Mosquito, 117–18; out-of-phase waves and, 19; of "squawk box," 29; of urban disturbance, 147

Audio Spotlight, 122, 147–48

automobiles, 13, 38, 47

aversive technology. *See* sound aversion technology

Aynsley-Green, Albert, 118

Baader-Meinhof Gang. *See* Red Army Faction

babies, 119; crying of, 85, 86, 87, 102

bagpipes, 91

Bain, Mark, 153

Banshee II, 114

*Barney*, 87, 136

barriers. *See* architectural barriers; "area denial"; electric fences; light barriers and fences; sound barriers and fences

battlefield simulation. *See* combat simulation

Battle of Jericho (Bible story), 91

behavior modification. *See* psychological control

Belgium, 119, 182n23

Bell Labs, 97, 98

Bess, Hellyette, 81

birds, 53, 92, 117, 122, 128

blasts, 16–17, 50, 55–56, 58. *See also* "acoustic blasters"; explosions

border control, 112, 148–50

Bosnia-Herzegovina, 100

brain, 8, 26, 40, 123, 124; damage/injury, 50, 124

Branch Davidian siege, Waco, 1993. *See* Waco siege, 1993

Bricet des Vallons, Georges-Henri, 129

Britain. *See* Great Britain

Broner, Nick, 22, 28, 30

"brown noise." *See* defecation, involuntary

B'Tselem, 81

buildings, 25, 41, 107, 148, 153

bullroarers, 39, 40

Bureau of Alcohol, Tobacco, Firearms, and Explosives, 60

Burris-Meyer, Harold, 96–97, 98, 105, 140–41, 174n21

Burroughs, William S., 27, 38, 152–53, 155n5

Bush, George H. W., 101

Bybee, Jay, 84

*Bylina*, 92–93

Bzorgi, Lee, 114

Cameron, Donald Ewen, 72, 76–77

camouflage and deception. *See* deception

Canada, 48, 68, 69–70, 77, 99, 112–13, 129, 181n2

cannons. *See* "air cannons"; "detonation engines"; "scare cannons"; "shock wave cannons"; vortex weapons

cardiac arrest. *See* heart attack

carnyxes, 91

cars. *See* automobiles

Catholic Church, 101

cell phones, 140

Celts, 91

Central America, 72

Central Intelligence Agency (CIA), 49–50, 67–68, 71–73, 76–78, 84, 86, 104, 127

Centre national de la recherche scientifique (CNRS), 24–28

Cerwin-Vega, 41

CETC International, 115, 149

chemical weapons, 46, 47, 59, 141, 163n19

children, 56, 103, 113, 119

China, 31, 67, 82–83, 90, 91–92, 112, 115, 149

Chion, Michel, 7

church organs. *See* pipe organs

civilians and noncombatants, 55–57, 130, 161n78

classical music, 99, 103, 118

Cline, Ray, 138

Coast Guard, U.S. *See* U.S. Coast Guard

Cold War, 67

"collateral damage," 60–61, 63–64, 65

collectives and collectivity, 2, 55, 145–46, 152–53, 154

colonial insurgency and anti-insurgency, 78–79, 93, 98–99, 128, 129

combat simulation, 142–44

Commission nationale de déontologie de la sécurité (CNDS), 64

communication, 8, 11, 21; isolation from (prevention of), 20, 101; technologies, 93, 111, 112, 121, 122–23, 130, 131, 181n2. *See also* propaganda

complex tones, 9–10

composers, 40, 43, 141

Compound Security Systems (CSS), 117–18

compressed-air pistols, 141

compressed-air sirens, 32, 33–34

concerts. *See* musical performances

Coppola, Francis Ford: *Apocalypse Now*, 99, 103

counterpiracy technology. *See* antipiracy technology

Coupland, Robin, 134

court cases and lawsuits: Canada, 77, 113; Europe, 79, 80; France, 119; Israel, 56–57, 82

Cox, Christoph, 96

criminal use, 132

crowd control, 35, 46, 47, 51, 52, 107–8, 129; at protests, 57–58, 63–64, 111, 112–13, 116. *See also* riot control

culturally offensive music, 88, 103, 104

Curdler, 106–7

Cusick, Suzanne, 88, 104, 135–36, 136–37

Davey Bickford, 63

David, Mike, 146, 148

Davison, Neil, 47–48; *"Non-Lethal" Weapons*, 5, 30, 145

deafness. *See* hearing loss

Decca, 141

deception, 95–96, 97, 100, 104, 122, 130, 133, 153

decibel, 13–16, 22, 156–57n18. *See also* amplitude

defecation, involuntary, 15, 23, 32

Defense Advanced Research Projects Agency (DARPA), 32, 34, 122, 129, 142

Defense Nuclear Agency, 52, 163n34

Defense Technologies, 61

demonstrations and protests. *See* protests and demonstrations

Department for Environment, Food and Rural Affairs (UK), 22

Department of Defense. *See* U.S. Department of Defense

depth-sounding, 51

Desert Storm. *See* Operation Desert Storm

detection and prospecting, 28–29, 115, 141, 142, 149, 164n49. *See also* sonar

detention. *See* prison and prisoners

deterrent devices. *See* sound aversion technology

"detonation engines," 43, 51, 52, 53–54, 151

Diamond Laboratories. *See* Harry Diamond Laboratories

Disney. *See* Walt Disney Company

Disperse, 107

disposable speaker systems, 105

dive bombers, 95, 97

divers, 36–37, 45, 52, 115

dogs, 9, 26, 117, 128, 157n30;
    barking of, 97, 153; to
    intimidate detainees, 84
drones, 105
drugs, 68, 76
Dunn, David, 13

ear, 1, 7–8, 11, 12, 13–14, 15;
    filtering by, 21; source detection
    by, 20, 157n30
EAR (U.S. Army). *See* Environment
    for Auditory Research (EAR)
eardrum, 8, 15, 25, 28, 44
earplugs, 69, 110, 116, 119
*Earthquake* (film), 41
earthquakes, 40, 55
echo, 17–18. *See also* sonar
Edgewood Research, Development
    and Engineering Center, 59
Egypt, 84, 96
E-Labs, 60
electric fences, 128
electric guitars, 87–88
electric shocks, 34, 68, 72, 77, 128.
    *See also* Taser
Electro-Optics Research &
    Development Ltd. (EORD),
    116, 148
EMI, 141
"enhanced interrogation," 83–90
Ensslin, Gudrun, 74
entertainment industry, 95, 97, 121,
    124, 125, 139, 140–44. *See also*
    films and film industry
Environment for Auditory Research
    (EAR), 142
equilibrium, 7, 8, 44, 106, 132
Ernst-Berendt, Joachim, 20
Escoitar, 1, 154
ethics, 64, 71, 90
Ethiopia, 99
Etienne Lacroix Group, 62
euphemism, 5, 84, 135, 139, 173n12
European Convention on Human
    Rights, 79, 80
European Union, 80, 120, 129, 149
experimentation on animals. *See*
    animal experimentation
experimentation on humans. *See*
    human experimentation
explosions, 14, 16–17, 49, 52–53,

151; underwater, 37, 51. *See also*
    blasts
explosive weapons, 43–63

Fairbanks, Douglas, Jr., 96, 97
Fallujah siege, 2004, 103–4, 112
Falun Gong, 82–83
fantasy weapons, 3, 27, 29–31, 32–33,
    138, 151. *See also* prototypes
fear, 30, 41, 79, 84, 146, 147. *See also*
    intimidation
Federal Bureau of Investigation
    (FBI), 61, 101–2
Federal Bureau of Prisons, 60
fences, electric. *See* electric fences
fences, light. *See* light barriers and
    fences
fences, sound. *See* sound barriers
    and fences
films and film industry, 41–42, 98,
    140
firecrackers, 43, 50
fireworks, 62
Flash-Ball, 133, 145
flash-bang grenades, 58–62, 63, 101,
    126, 165n81
flashes, 43, 52, 58
Fletcher, Harvey, 97
food deprivation, 67, 80, 83
Foucault, Michael, 126
France, 57–58, 62–64, 81, 98, 116–
    17, 119; prisoners, 67, 73–75;
    research centers and institutes,
    24–28, 123
Franks, Tommy, 100
Fridman, Igor, 54
Friedman, Herbert, 112

Galen, 21
García López, Noel, 145, 147
Garnier, Jean-Pierre, 150
gas weapons. *See* chemical weapons
Gavreau, Vladimir, 24–28, 38
Gayl blaster, 34, 108, 160n58
Gaza Strip, 55–57
Geneva Conventions, 84
*German Research in World War II*
    (Simon), 45–46
Germany, 45–46, 48–49, 58, 62, 95,
    96, 141
"ghost armies," 95–96, 140

ghosts and ghost sounds, 38–39, 86, 94, 99, 122
glass-breaking, 11–12
Global Strategy Council, 138
Goodman, Steve, 6, 24, 148, 150, 153–54
Goodrich, 60
Gottlieb, Sidney, 68
Grant, Steven, 143
Great Britain, 22, 29, 40–41, 47–48, 58, 96; army, 62, 91; CIA collaboration, 68, 78; in colonies, 93, 98; military-entertainment complex, 141; military interrogation techniques, 78–81; Ministry of Defense, 30, 35, 45, 48; Mosquito in, 118–19, 119–20; NIJ cooperation agreement, 129; riot control, 29–30, 107
Grenada, 100
grenade launchers, 47, 62
grenades, 2, 57–65, 101, 126, 144, 151. See also flash-bang grenades
Grimes, Joseph, 143–44
Grisham, C.J., 87
Gross, Jan, 73
Gruenler, Carl, 112
GT Devices, 52
G20 summits, 111, 112–13, 116
Guantánamo Bay, Cuba, 84, 85–86, 88–89, 135–36
guinea pigs, human. See human experimentation

Hadsell, Mark, 88
hailing devices, 105, 114, 145. See also long-range acoustic devices (LRAD)
Haiti, 100, 112
Hall, Wade, 114
hallucinations, 70, 71, 72, 73, 78, 81, 103
Hambling, David, 52–53
harassment, 87, 94, 99–100, 103–4
Harbor Subsurface Protection System, 37
harmonic frequencies, 9
Harris, Stanley, 28
Harry Diamond Laboratories, 107
"haunted" laboratories, 38–39

headache, 15, 23, 26, 61, 70, 111, 119
hearing, collective nature of, 145–46
hearing loss, 14, 34, 44, 56, 57, 61, 103, 106, 145; from LRAD, 111, 113
heart attack, 44, 60
heavy metal music, 87–88, 101, 103, 136
Hebb, Donald O., 69–70, 76, 77, 86
helicopters, 33, 99–100, 102
Helms, Richard, 68
Hergé: Tintin series, 11, 49
Hetfield, James, 136
high-frequency sounds, 2, 8, 9; audibility, 9, 11; in deterrence, 115, 117–20, 124, 160n58; diffraction, 17; infrasound production and, 29; physiological effects, 15, 106
high-frequency weapons, 3, 21, 33, 65, 93–94, 105–6, 124, 152
hijackers, 58, 109
Hills, James T., 84
hissing, 50–51, 80, 123
Holosonics, 121, 122
homeless people, control of, 117–18, 122, 184n66
"Honduran Manual." See Human Resource Exploitation Training Manual
Honduras, 112
hooding, 80, 81–82, 83, 84, 85
horns and trumpets. See trumpets
hostage rescue, 32, 34, 35, 60, 122, 123
HPS-1, 30, 99, 106
HPV Technologies, 114
human experimentation, 35–36, 40–41, 89–90, 116, 128; "non-volunteer," 71, 76–77, 128; sensory deprivation, 69–71, 73, 78; underwater, 52
Human Resource Exploitation Training Manual, 73, 85
Hungary, 30

Ilahita Arapesh, 39
imagination, 24, 42, 137–40, 146. See also fantasy weapons
IMLCORP, 114
Indusec, 115, 179n113
infants. See babies

Inferno Intenso, 115, 147
"infrasonic fish fence," 38
infrasonic weapons, 2, 25–38, 52
infrasounds, 12, 21–42, 151;
    audibility, 8–9, 156n8; in
    automobiles, 13, 38; in crowd
    control, 107–8; in films and
    music, 27, 40–42; physiological
    effects, 11, 39, 42, 159n31; for
    probing and localization, 29;
    propagation, 20; psychological
    effects, 22, 38, 40, 41
insomnia. See sleeplessness
Institute for Non-Lethal Defense
    Technologies, 146
international protocol and law, 65,
    82, 84, 133, 134, 161n78, 171n93
interrogation, 71–73, 78–90, 127,
    170n73, 171n93
intestines, 11, 15, 22, 33. See also
    defecation, involuntary
intimidation, 55–57, 79–80, 84, 92,
    93, 113
IPB, 119
Iraq, 84, 89, 100, 103–5, 112, 135,
    136, 171n93, 172n1
Ireland, 79, 119. See also Northern
    Ireland
Irréversible (film), 41–42
isolation, 67, 73–76, 77, 80, 81, 84.
    See also solitary confinement
Israel, 53–57, 81–82, 90, 115–16, 129

Jamaica, 93
Japan, 29, 95, 97, 111
Jericho, Battle of. See Battle of
    Jericho (Bible story)
Johnson, Daniel, 28
Joint Non-Lethal Weapons
    Directorate (JNLWD), 33, 47,
    53, 59, 114, 115
Jollivet, Boris, 13

Kintner, Del, 110
kites, 91–92
Kittler, Friedrich, 140, 141
Korean War, 98
KUBARK Counterintelligence
    Interrogation, 71–72, 85
Kuehl, Dan, 104

Laboratoire de mécanique et
    d'acoustique (LMA), 24–28
Lacroix-Ruggieri, 62
The Lancet, 78
land mines, 32, 109, 149
laser weapons, 3, 50, 52–53, 65, 130,
    145, 149
law enforcement, 1, 6, 105, 129. See
    also police
Lawrence, Tony R., 38, 40
laws and regulations, 15, 64–65, 119,
    129. See also court cases and
    lawsuits; international protocol
    and law
Lenoir, Timothy, 142
lethality, 28, 30, 44, 54–55, 108,
    124, 138; of grenades, 58, 60; of
    LRADs, 113; as part of a range,
    36, 131; of torture, 72. See also
    "non-lethality"
Levavasseur, Robert, 158
Levy, Gideon, 55
Lewer, Nick, 47–48
Liénard, Pierre, 24, 26, 92
light: in crowd control, 107–8;
    deprivation of, 72, 84; legal
    aspects, 119; measurement of,
    165n71; as weapon or method of
    torture, 30, 83, 84, 101, 113. See
    also flash-bang grenades; laser
    weapons
light barriers and fences, 179n113
Lilly, John C., 70–72, 168n19
LMA. See Laboratoire de mécanique
    et d'acoustique (LMA)
Lockheed Martin, 37
loitering prevention, 118, 148
Longina, Chiu, 154
long-range acoustic devices
    (LRADs), 15–16, 53, 104–6,
    110–14, 131, 145; for animal
    control, 128; bloodlessness of,
    133; in combat simulation, 143;
    hybrid/dual use, 104, 110, 121;
    police use, 111, 112–13, 130,
    181n2; private use, 132
Loos, Hendricus G., 159n31
López, Juan-Gil, 154
Los Angeles, 122, 129, 148
loudness. See amplitude
loudspeakers, 20, 33, 34, 53, 94–110.
    See also long-range acoustic

devices (LRAD); subwoofers; ultrasonic loudspeakers
low-frequency sounds, 8, 9, 11, 21; in deterrence, 35–36, 37; in films, 41–42; of Gavreau device, 26; industrial, 22; in LRADs, 114; masking effect, 16; physiological effect, 12, 15, 22, 23–24, 28; propagation, 17, 20; of sirens, 38; of subwoofers, 156n11
low-frequency weapons, 3, 29, 33, 34, 35, 43–44, 115–16
LRAD Corp., 109–11, 112, 121, 122
LRADs. See long-range acoustic devices (LRADs)
lungs, 12–13, 15, 28, 44

Marine Corps. See U.S. Marine Corps
marketing, subliminal, 121–22
Marks, John, 67, 68
Martin, Brian, 149
Mayer, Jane, 85
McCoy, Alfred: A Question of Torture, 68, 78, 90
McGill University, 69, 76–77
media, 127, 130, 135, 139, 173n13
medical use of sound, 13, 156n14, 159n40
medium-frequency sounds, 9, 22, 38, 160n58
medium-frequency weapons, 21, 33, 65, 91–124, 152
MEDUSA, 123
megaphones, 104, 113, 181n2
Meinhof, Ulrike, 74–75
Meins, Holger, 75
Menigon, Nathalie, 67
Metallica, 87, 101, 136
microphones, 8, 20, 101, 110
"microwave hearing," 123
microwaves, 6, 113, 149–50
military-entertainment complex, 95, 124, 139, 140–44
military-industrial complex, 130, 142
military research, 4, 23–31, 53–54, 78, 141; United States, 31–38, 46–47, 50–53, 59, 107–9, 113–14, 127–28, 142
Miller, General Geoffrey, 85

mind control. See psychological control
mines. See land mines
Missouri University of Science and Technology, 143
mistral, 21
MKUltra. See Project MKUltra
Mobile Acoustic System (AFRL), 34
mobile apps, 140
Mohamed, Binyamin, 86–87
Mohr, George C., 23
Morocco, 84, 86
Morris, Chris, 138
Morris, Janet, 31, 138
Morse code, 123, 141
Mosquito ringtone, 140
Mosquito (sound aversion device), 117–20, 131, 133, 147, 182n23
Mullins, Justin, 50–51
music, 6, 10, 27; anti-adolescent, 118; in military operations, 91, 93, 94, 98, 99, 103–4, 104, 136; in sieges, 102, 112; as torture, 2, 3, 66, 82, 83, 84, 85–86, 87–89, 90, 103, 136–37. See also classical music; heavy metal music; rap music
musical instruments, 9, 39, 41, 91, 93. See also trumpets
musical performances, 17–18, 41
music recording industry, 141
Muslims, 87, 103
Muzak, 96, 118, 148, 173n19
myths and false claims about sound, 14–15, 21, 22, 24–28, 33. See also fantasy weapons

National Aeronautics and Space Administration (NASA), 23, 29
National Defense Research Council (NDRC), 97
National Institute of Justice (NIJ), 34, 59, 60, 61, 129
National Institute of Mental Health (NIMH), 70–71
National Nuclear Security Administration, 61
National Security Technology Center, 114
National Tactical Officers Association, 60
nausea: as effect of silence, 70; as

effect of sound, 15, 22, 23, 115, 119, 149
Nazis, 2, 45–46, 48–49, 95, 96, 141
Netherlands, 75
New Guinea, 39
New York City, 60, 99, 111, 121
Noé, Gaspar, 41
noise, 10, 43, 65, 152; as torture, 80, 85, 87. *See also* anti-noise headsets and helmets; anti-noise regulations
noncombatants and civilians. *See* civilians and noncombatants
Non-Lethal Acoustic Weapons Program, 33
"non-lethality," 2, 4, 5, 33, 125, 126–34
Non-Lethal Swimmer Neutralization Study, 37
Non-Lethal Technologies, 165n81
*"Non-Lethal" Weapons* (Davison), 5
Noriega, Manuel, 101, 103
Norris, Elwood, 110, 111, 121
Northern Ireland, 29, 79, 107
"no-touch torture." *See* torture: psychological
nuclear site protection, 32, 115, 129

obedience, 146, 147
obstacles to sound, 17, 18, 20, 120; circumvention, 54
Occupational Safety and Health Administration (OSHA), 143
octave, 9
Office of Naval Research (ONR), 50, 68, 159n40
Office of Public Safety, 72
opera, 101
Operation Demetrius, 79
Operation Desert Storm, 100
Operation Fortitude, 95–96
Operation Just Cause, 100–101

Page, Jimmy, 27
pain: auditory, 30, 55, 59, 64–65, 108; as effect of infrasounds, 25, 26; as effect of low-frequency sounds, 23, 26; inflicted by LRAD, 113; in interrogation, 81; threshold, 14, 22, 64–65
Palestinians, 55–57, 79, 81–82, 90, 116

Panama, 100–101
panic, 34, 45, 56, 78, 95, 97, 107
panopticon, 148
parabolic generators, 32, 48–49
PDT Agro, 53–54
*Petite histoire de l'acoustique* (Liénard), 24, 26, 92
Phantom jets, 55
phones, cell. *See* cell phones
Physicians for Human Rights, 89
Physicians for Human Rights Israel, 56
physiological effects of silence, 73
physiological effects of sound, 3, 11, 12, 15, 22, 23, 44, 104; alleged/"overstated," 15, 23, 30–31, 32, 33, 34–35, 37, 108, 115–16; of sonic booms, 88, 56. *See also* hearing loss; tinnitus
Picatinny Arsenal, 32, 35, 53
Pieslak, Jonathan, 87, 103
piezoelectricity, 177n89
pipe organs, 41
piracy deterrence. *See* antipiracy technology
pistols, compressed-air. *See* compressed-air pistols
Pittsburgh, 111
"plasma" weapons, 50–53, 64, 151
polar explorers, 71
police, 35, 118, 126; interrogation training and techniques, 67, 72–73; France, 62, 63–64; HPS-1 use, 99; Los Angeles, 122; loudspeaker use, 99, 111–12, 116, 130, 181n2; Northern Ireland, 79; public relations, 138; self-policing, 134; sounds of, 38, 93, 106, 158n22
political prisoners, 67, 73–75, 79–80, 81–83, 169n35
Pompei, Joseph, 121
Porter, Catherine, 181n2
port protection, 52
Powell, Colin, 101
Power Sonix, 105
"pre-lethal" weapons, 94, 112, 124, 144, 152
Primex Physica International, 52
prisons and prisoners: China, 82–83; France, 67; Germany, 73–75; Great Britain, 79–80; grenades

prisons and prisoners (*continued*) and, 60, 144; Guantánamo Bay, 84, 85–86, 88–89, 135–36; Iraq, 84, 86, 171n93; Israel, 81–82; Northern Ireland, 79; POWs, 72; sensory deprivation/solitary confinement, 71, 72, 73–75, 79; United States, 60, 61, 83–90, 112; "virtual fences" and, 148. *See also* interrogation; political prisoners

private use, 117, 119, 124, 131–32, 150, 182n23

Project MKUltra, 49–50, 68, 71, 77, 83, 102, 127

Project Sheriff, 113–14

Project Sirius, 134

proliferation, 129–32

Proll, Astrid, 74, 75

propaganda, 94, 95, 98–99, 105, 139–40

propane explosions, 53

proportionality, 65, 132

prospecting and detection. *See* detection and prospecting

protests and demonstrations, 126, 146; use of aversive sound against, 107, 111, 112–13, 116; use of grenades against, 57–58, 63–64

prototypes, 26, 30, 33, 35, 37, 46, 52, 110, 151; LRADs, 113; underwater, 52, 115

"psycho-galvanometer," 141

psychological control, 49, 66, 67, 114–15, 137–40. *See also* torture: psychological

psychological operations (psyops), 94–124, 127, 173n12–13

psychological weapons, 51, 55–56, 58, 63, 66

public health codes, 119

public perception and opinion, 90, 127, 135–36, 139

public space, control of, 145–50

pulse generators, 33, 34, 54

pure tones, 9, 109, 110

pyrotechnics. *See* fireworks

questioning of detainees. *See* interrogation

A Question of Torture (McCoy), 68

radio frequencies, 123

Railey, Hilton Howell, 96, 97

rap music, 86, 88, 103

Rappert, Brian, 134

Raytheon, 113

Razac, Olivier, 125, 144

recording industry. *See* music recording industry

Red Army Faction, 73–75

refugees, 149

regulations. *See* laws and regulations

religious use of sound, 39, 40, 41

resistance, collective, 151, 152–53, 154

resonance frequency, 11–12

reverberation. *See* echo

Rheinmetall Defence, 62

"rheostatic weapons," 131

"Ride of the Valkyries" (Wagner), 99, 103

riot control, 29–30, 54, 57, 107, 123

*Riot Control* (Applegate), 29, 57

Roads, Curtis, 43

Roman Catholic Church. *See* Catholic Church

Rote Armee Fraktion. *See* Red Army Faction

rubber shrapnel, 62, 63

Rubinstein, Leonard, 77

Ruggieri, 62

"Rumblers" (sirens), 38

rumor, 31, 95, 139, 151

Russia, 31, 48, 61–62, 115

Rutledge, Bill, 100

SADAG. *See* Sequential Arc Discharge Acoustic Generator (SADAG)

Sanchez, Ricardo, 84

Sandia National Laboratories, 59, 61

SARA. *See* Scientific Applications and Research Associates (SARA)

Sarkozy, Nicolas, 63

Saul, Henri, 26

"scare cannons," 92, 128

Schafer, Murray, 93, 152

Schattle, Duane, 135

Scheuer, James, 126

School of the Americas, 72

Schweiker, Richard, 50

science fiction, 42, 138, 141

Scientific Applications and Research Associates (SARA), 32–33, 34, 48, 59, 105, 108, 131
*The Search for the "Manchurian Candidate"* (Marks), 67
Sea Shepherd, 111, 132
secrecy, 4, 50, 137–40, 139
*Secret Weapons of the Third Reich* (Simon), 49, 162n8
Seebeck, Thomas Johann, 93
seismic prospecting, 164n49
self-defense, 35, 64, 131–32
sensational claims. *See* myths and false claims about sound
sensory deprivation, 66–67, 69–76, 83, 127
sensory overload, 59, 102. *See also* sound saturation
Sensurround, 41
Sequential Arc Discharge Acoustic Generator (SADAG), 34, 35
SERE. *See* Survival, Evasion, Resistance and Escape (SERE)
Shelton, Thomas, 46, 141
Sheriff Project. *See* Project Sheriff
ships, 105, 111, 114, 122, 132, 141
"shock wave cannons," 53–55, 64
shock waves, 16–17, 43, 44, 49, 50–51, 59; of grenades, 61, 63; possible lethality, 27, 44; in seismic prospecting, 164n49
shopping malls, 118
shouting, 82, 84, 89, 91; mechanized, 115
shrapnel, 43, 62, 63
silence, 6, 67; as effect of torture, 137; need for, 14; physiological effects, 73; psychological effects, 69–70; as torture, 73–76
Simon, Leslie E., 45–46, 49, 162n8
Singleton, Bob, 136
sirens, 38, 95, 97, 104, 108, 115, 116, 160n58; amplitude, 106; invention, 93. *See also* compressed-air sirens
Sirius Project. *See* Project Sirius
Sironi, Françoise, 137
sleep deprivation, 67, 80, 81–82, 83, 88, 102, 168n24
sleeplessness, 56, 61, 73, 159n31
SMBI. *See* Stress and Motivated Behavior Institute (SMBI)

smoke bombs and smoke screens, 47, 59, 104
Society for the Investigation of Human Ecology, 77
Somalia, 58, 100, 111
sonar, 6, 18, 37, 51, 159n40
sonic booms, 54, 55–57
sonic bullet. *See* "acoustic bullet"
sonic detection and prospecting. *See* detection and prospecting
sonic fences. *See* sound barriers and fences
Sonic Grenade (phone app), 140
*Sonic Weapons* (Escoitar), 154
sound: absorption and diffraction, 17, 20; definition of, 7, 156n3; directionality, 20, 157n30; discomfort threshold, 14, 106, 120; extra-auditive effects, 11; frequency, 8–13, 14; masking effect, 16; propagation, 7, 16–20, 36, 40, 156n3; spectrum, 6, 8–9. *See also* high-frequency sounds; low-frequency sounds; medium-frequency sounds
sound, medical use. *See* medical use of sound
sound, physiological effects. *See* physiological effects of sound
sound, religious use. *See* religious use of sound
sound aversion technology, 105–20, 122, 125, 128, 149
sound barrier breaking, 5, 16–17. *See also* sonic booms
sound barriers and fences, 32, 38, 101, 104, 114, 147, 148
sound-blocking devices. *See* anti-noise headsets and helmets; earplugs
"sound cannons." *See* long-range acoustic devices (LRAD)
SoundCommander, 114
sound effects, 96, 102. *See also* ghost sounds
soundproof rooms and cells, 17, 70, 72, 73, 78, 82, 167n14
sound repellant technology. *See* sound aversion technology
sound reverberation. *See* echo
sound saturation, 76–90, 102

sound signatures. *See* acoustic
    signatures
"sound torpedo," 98, 105
sound volume. *See* amplitude
sound waves, 7, 9–11, 18–20
South America, 72
Soviet Union, 31, 46, 67, 138
space denial. *See* "area denial"
speed of sound, 16–17
Speer, Albert, 45
"squawk box," 29–30, 35, 116
Stapleton, Howard, 117
Stellar Photonics, 52
stereo sound systems, 20, 95, 97,
    98, 140
Stiner, Carl, 101
Stockhausen, Karlheinz, 141
Stone, Alan, 102–3
Stress and Motivated Behavior
    Institute (SMBI), 36, 109, 128
stroboscopes, 30, 179n113
"stun grenades," 53, 60
subliminal messages, 102, 120–23
subliminal sounds. *See* infrasounds;
    ultrasounds
submarines and submariners, 28, 98
subwoofers, 11, 34, 40, 42, 156n11
suicide, 75
Sun Tzu, 127
supersonic boom. *See* sonic boom
surveillance, 122, 148, 149
Survival, Evasion, Resistance and
    Escape (SERE), 85
Sweden, 75, 115
swimmer deterrence, 36, 37, 52, 115
Synetics Corporation, 35

Tamabaran cult, 39
Tandy, Vic, 38–39, 40
Target Behavioral Response
    Laboratory (TBRL), 36, 127–28
Taser, 63, 128, 149
teargas, 63
teenagers, 117–20
temporary threshold shift (TTS), 14
terror: creation of, 55–56, 59; war
    on, 2, 56, 60, 78, 84, 90, 128,
    150
Teuns, Sjef, 75–76
Thailand, 111
Thomas, Timothy L., 132
Throbbing Gristle, 27

timbre, 9
tinnitus, 15, 44, 61, 119, 157n26
Tintin, 11, 49
Toffler, Alvin, 138
Toffler, Heidi, 138
tornadoes, 45
torture, 66, 67, 80, 81–90, 132,
    151–52, 168n24, 170n73;
    psychological, 68, 71, 72–90,
    135–36. *See also* music: as
    torture
toy guns, 141
trumpets, 91, 93, 116, 156n10
Turner, Stansfield, 50
Tuzin, Donald, 39
tympanic membrane. *See* eardrum

ultrasonic deterrent devices, 107–8,
    117, 128
ultrasonic imaging, 156n14
ultrasonic loudspeakers, 120–22,
    131, 144
ultrasonic weapons, 32, 35
ultrasounds, 9, 11, 156n8, 159n40
underwater detection devices, 141
underwater weapons and deterrent
    devices, 32, 36–38, 44–45, 51–
    52, 115, 117
United Kingdom. *See* Great Britain
United Nations, 56, 57, 82, 139,
    163n19
United States, 58–61, 83–91;
    military research in, 31–38, 46–
    47, 50–53, 59, 107–9, 113–14,
    127–28, 142; psyops, 94–105.
    *See also* Central Intelligence
    Agency (CIA)
Universal Propulsion Company,
    59–60
universities, 37, 51, 73, 69, 76–77,
    107, 113, 143
"urbanism of sound," 144, 145–50
U.S. Air Force, 33, 34
U.S. Army, 33–34, 46–47, 50–51,
    52–53, 59, 60; interrogation
    techniques, 84, 171n93; on
    "microwave hearing," 123;
    psyops, 95, 96, 97, 99, 101, 105,
    122, 127; sound equipment of,
    30, 105, 141; Special Operations
    Command, 105. *See also*
    Army Armament Research,

Development and Engineering Center (ARDEC); Army Research Laboratory (ARL)
U.S. Coast Guard, 61, 115
U.S. Department of Defense: *Field Manual 34-52*, 89, 171n93
U.S. Marine Corps, 59
U.S. Navy, 50, 52, 68, 96, 105, 122, 123, 159n40
USSR. *See* Soviet Union

Vance, Donald, 89
video games, 5, 140, 142
Vietnam War, 30, 72, 73, 99, 106
Villatoux, Marie-Catherine, 95
Villatoux, Paul, 95
Vinokur, Roman, 92–93
"virtual fences." *See* light barriers and fences; sound barriers and fences
*Voices from Heaven* (Villatoux and Villatoux), 95
volume (loudness). *See* amplitude
vortex weapons, 31, 45–47, 53, 64, 141, 151

Waco siege, 1993, 101–2, 129
Wagner, Richard: "Ride of the Valkyries," 99, 103
Wallauscheck, Richard, 48–49
Walt Disney Company, 97, 121, 140, 150
war, 130–31, 132–33, 142–44; international law, 161n78. *See also* Afghanistan; Iraq; Vietnam War; World War I; World War II
warning devices, 114–17. *See also* alarms; sirens
water, sound travel through, 17, 36. *See also* underwater detection devices; underwater weapons and deterrent devices

water torture, 82, 84, 85
Wattre Corporation, 114
weapons, antiaircraft. *See* antiaircraft weapons
weapons, explosive. *See* explosive weapons
weapons, fantasy. *See* fantasy weapons
weapons, high-frequency. *See* high-frequency weapons
weapons, laser. *See* laser weapons
weapons, low-frequency. *See* low-frequency weapons
weapons, medium-frequency. *See* medium-frequency weapons
weapons, "plasma." *See* "plasma" weapons
weapons, pre-lethal. *See* "pre-lethal" weapons
weapons, psychological. *See* psychological weapons
weapons, vortex. *See* vortex weapons
"weapons of mass destruction," 30, 33
Weidlinger Associates, 37
Westervelt, Peter, 159n40
whistles and whistling sounds, 26, 79, 92–93, 158n22
Whiteley, Sheila, 87–88
white noise, 10
wind turbines, 21–22
World War I, 29
World War II, 2, 29, 45–46, 66, 93, 94–98, 127, 141
Wortman, Andrew, 46
Wright, Steve, 128–29, 133, 149
Wyle Laboratories, 35

yelling. *See* shouting
Yoo, John, 84

Zero dB, 89
"zones of exclusion." *See* area denial

# PUBLISHING IN THE PUBLIC INTEREST

Thank you for reading this book published by The New Press. The New Press is a nonprofit, public interest publisher. New Press books and authors play a crucial role in sparking conversations about the key political and social issues of our day.

We hope you enjoyed this book and that you will stay in touch with the New Press. Here are a few ways to stay up to date with our books, events, and the issues we cover:

- Sign up at www.thenewpress.com/subscribe to receive updates on New Press authors and issues and to be notified about local events
- Like us on Facebook: www.facebook.com/newpress books
- Follow us on Twitter: www.twitter.com/thenew press

Please consider buying New Press books for yourself; for friends and family; or to donate to schools, libraries, community centers, prison libraries, and other organizations involved with the issues our authors write about.

The New Press is a 501(c)(3) nonprofit organization. You can also support our work with a tax-deductible gift by visiting www.thenewpress.com/donate.